Handbook of Materials Structures, Properties, Processing and Performance

Handbook of Materials Structures, Properties, Processing and Performance

Editor

Yogesh Rathod

Handbook of Materials Structures, Properties, Processing and Performance

Edited by **Yogesh Rathod**

Printed in 2017

ISBN: 978-1-68117-220-0

Library of Congress Control Number: 2015936580

© 2016 by
SCITUS Academics LLC,
616, Corporate Way, Suite 2, 4766,
Valley Cottage, NY 10989

www.scitusacademics.com

This book contains information obtained from highly regarded resources. Copyright for individual articles remains with the authors as indicated. All chapters are distributed under the terms of the Creative Commons Attribution License, which permits unrestricted use, distribution, and reproduction in any medium, provided the original author and source are credited.

Notice

Reasonable efforts have been made to publish reliable data and views articulated in the chapters are those of the individual contributors, and not necessarily those of the editors or publishers. Editors or publishers are not responsible for the accuracy of the information in the published chapters or consequences of their use. The publisher believes no responsibility for any damage or grievance to the persons or property arising out of the use of any materials, instructions, methods or thoughts in the book. The editors and the publisher have attempted to trace the copyright holders of all material reproduced in this publication and apologize to copyright holders if permission has not been obtained. If any copyright holder has not been acknowledged, please write to us so we may rectify.

Contents

Preface vii

Chapter 1	Properties of Concrete at Elevated Temperatures 1
	Venkatesh Kodur

Chapter 2	Aerogels as Promising Thermal Insulating Materials: An Overview 43
	Prakash C, Thapliyal, and Kirti Singh

Chapter 3	Atmospheric Corrosion of Painted Galvanized and 55%Al-Zn Steel Sheets: Results of 12 Years of Exposure 71
	C. I. Elsner, P. R. Seré, and A. R. Di Sarli

Chapter 4	The Addition of Graphene to Polymer Coatings for Improved Weathering 109
	Nurxat Nuraje, Shifath I. Khan, Heath Misak, and Ramazan Asmatulu

Chapter 5	Recent Progress in Processing of Tungsten Heavy Alloys 129
	Y. Şahin

Chapter 6	Synthesis of β-SiC Fine Fibers by the Forcespinning Method with Microwave Irradiation 189
	Alfonso Salinas, Maricela Lizcano, and Karen Lozano

Chapter 7	Foaming Behaviour, Structure, and Properties of Polypropylene Nanocomposites Foams 203
	M. Antunes, V. Realinho, and J. I. Velasco

Chapter 8	Natural Products: A Minefield of Biomaterials 235
	Oladeji O. Ige, Lasisi E. Umoru, and Sunday Aribo

Citations 291

Index 295

Preface

The book incorporates a historical account of critical developments and the evolution of materials fundamentals, providing an important perspective for materials innovations, including advances in processing, selection, characterization, and service life prediction. It includes the perspectives of materials chemistry, materials physics, engineering design, and biological materials as these relate to crystals, crystal defects, and natural and biological materials hierarchies, from the atomic and molecular to the macroscopic, and emphasizing natural and man-made composites. Dr. Murr's expansive presentation of topics in Materials Properties and Performance: A Handbook for Engineers & Scientists explores inter-relationships among materials properties, processing, and synthesis (both historic and contemporary) while maintaining a highly readable "narrative" style and an encyclopedic breadth of coverage. The book serves as both an authoritative reference and roadmap of advanced materials concepts for practitioners, graduate-level students, and faculty coming from a range of disciplines.

Editor

Chapter 1

Properties of Concrete at Elevated Temperatures

Venkatesh Kodur

Department of Civil and Environmental Engineering, Michigan State University, East Lansing, MI 48824, USA

ABSTRACT

Fire response of concrete structural members is dependent on the thermal, mechanical, and deformation properties of concrete. These properties vary significantly with temperature and also depend on the composition and characteristics of concrete batch mix as well as heating rate and other environmental conditions. In this chapter, the key characteristics of concrete are outlined. The various properties that influence fire resistance performance, together with the role of these properties on fire resistance, are discussed. The variation of thermal, mechanical, deformation, and spalling properties with temperature for different types of concrete are presented.

INTRODUCTION

Concrete is widely used as a primary structural material in construction due to numerous advantages, such as strength, durability, ease of fabrication, and non-combustibility properties, it possesses over other construction materials. Concrete structural members when used in buildings have to satisfy appropriate fire safety requirements specified in building codes [1–4]. This is because fire represents one of the most severe environmental conditions to which structures may be subjected; therefore, provision of appropriate fire safety measures for structural members is an important aspect of building design.

Fire safety measures to structural members are measured in terms of fire resistance which is the duration during which a structural member exhibits resistance with respect to structural integrity, stability, and temperature transmission [5, 6]. Concrete generally provides the best fire resistance properties of any building material [7]. This excellent fire resistance is due to concrete's constituent materials (i.e., cement and aggregates) which, when chemically combined, form a material that is essentially inert and has low thermal conductivity, high heat capacity, and slower strength degradation with temperature. It is this slow rate of heat transfer and strength loss that enables concrete to act as an effective fire shield not only between adjacent spaces but also to protect itself from fire damage.

The behaviour of a concrete structural member exposed to fire is dependent, in part, on thermal, mechanical, and deformation properties of concrete of which the member is composed. Similar to other materials the thermophysical, mechanical, and deformation properties of concrete change substantially within the temperature range associated with building fires. These properties vary as a function of temperature and depend on the composition and characteristics of concrete. The strength of concrete has significant influence on its properties at both room and high temperatures. The properties of high strength concrete (HSC) vary differently with temperature than those of normal strength concrete (NSC). This variation is more pronounced for mechanical properties, which are affected by strength, moisture content, density, heating rate, amount of silica fume, and porosity.

In practice, fire resistance of structural members used to be evaluated mainly through standard fire tests [8]. In recent years, however, the use of numerical methods for the calculation of the fire resistance of structural members is gaining acceptance because these calculation methods are far less costly and time consuming [9]. When a structural member is subjected to a defined temperature-time exposure during a fire, this exposure will cause a predictable temperature distribution in the member. Increased temperatures cause deformations and property changes in the constitutive materials of a structural member. With knowledge of deformations and property changes, the usual methods of structural mechanics can be applied to predict the fire resistance performance of a structural member. The availability of material properties at an elevated temperature permits a mathematical approach for predicting fire resistance of structural members [10, 11].

Clearly, the generic information available on properties of concrete at room temperature is seldom applicable in fire resistance design [12]. It is imperative, therefore, that the fire safety practitioner knows how to extend, based on a priori considerations, the utility of the scanty property data that can be gathered from the technical literature. Also, knowledge of unique characteristics, such as fire induced spalling in concrete, is critical to determine the fire performance of concrete structural members.

PROPERTIES INFLUENCING FIRE RESISTANCE

General

The fire response of reinforced concrete (RC) members is influenced by the characteristics of constituent materials, namely, concrete and reinforcing steel. These include (a) thermal properties, (b) mechanical properties, (c) deformation properties, and (d) material specific characteristics such as spalling in concrete. The thermal properties determine the extent of heat transfer to the structural member, whereas the mechanical properties of constituent materials determine the extent of strength loss and stiffness deterioration of the member. The deformation properties, in conjunction with mechanical properties, determine the extent of deformations and strains in the structural member. In addition, fire induced spalling of concrete can play a significant role in the fire performance of RC members [13]. All these properties vary as a function of temperature and depend on the composition and characteristics of concrete as well as those of the reinforcing steel [12]. The temperature induced variation in properties in concrete is much more complex than that in reinforcing steel due to moisture migration as well as significant variation of ingredients in different types of concrete. Thus, the primary focus of this chapter is on the effect of temperature on properties of concrete. The effect of temperature on properties of steel reinforcement can be found elsewhere [4, 12].

Concrete is available in various forms and it is often grouped under different categories based on weight (as normal weight and light weight concrete), strength (as normal strength, high strength, and ultrahigh strength concrete), presence of fibers (as plain and fiber-reinforced concrete), and performance (as conventional and high performance concrete). Fire safety practitioners further subdivide normal-weight concretes into silicate (siliceous) and carbonate (limestone) aggregate concrete, according to the composition of the principal aggregate. Also, when a small amount of discontinuous fibers (steel or polypropylene) is added to a concrete batch mix to improve performance, this concrete is referred to as fiber-reinforced concrete (FRC). In this section, the various properties of concrete are mainly discussed for conventional concrete. The effect of strength, weight, and fibers on properties of concrete at elevated temperatures is highlighted.

Traditionally, the compressive strength of concrete used to be around 20 to 50 MPa, which is classified as normal-strength concrete (NSC). In recent years, concrete with a compressive strength in the range of 50 to 120 MPa has become widely available and is referred to as high-strength concrete (HSC). When compressive strength exceeds 120 MPa, it is often referred to as ultrahigh performance concrete (UHP). The strength of concrete degrades with temperature and the rate of strength degradation is highly influenced by the compressive strength of concrete.

Thermal Properties

The thermal properties that influence temperature rise and distribution in a concrete structural member are thermal conductivity, specific heat, thermal diffusivity, and mass loss.

Thermal conductivity is the property of a material to conduct heat. Concrete contains moisture in different forms, and the type and the amount of moisture have a significant influence on thermal conductivity. Thermal conductivity is usually measured by means of "steady state" or "transient" test methods [14]. Transient methods are preferred to measure thermal conductivity of moist concrete over steady-state methods [15–17], as physiochemical changes of concrete at higher temperatures because intermittent direction of heat flow On average, the thermal conductivity of conventional normal strength concrete, at room temperature, ranges between 1.4 and 3.6 W/m-°C [18].

Specific heat is the amount of heat per unit mass, required to change the temperature of a material by one degree and is often expressed in terms of thermal (heat) capacity which is the product of specific heat and density. Specific heat is highly influenced by moisture content, aggregate type, and density of concrete [19–21]. The variation of specific heat with temperature used to be determined through adiabatic calorimetry until 1980s. Since the 1980s, differential scanning calorimetry (DSC) has been the most commonly used technique for mapping the curve in a single temperature sweep at a desired rate of heating [22, 23]. Unfortunately, the accuracy of the DSC technique in determining the sensible heat contribution to the apparent specific heat may not be particularly good (sometimes it may be as low as ±20 percent). The rate of temperature rise in DSC tests is usually $5°C \cdot min^{-1}$. At higher heating rates, the peaks in the DSC curves tend to shift to higher temperatures and become sharper. For temperatures above 600°C, a high-temperature differential thermal analyzer (DTA) is also used to evaluate specific heat.

The thermal diffusivity of a material is defined as the ratio of thermal conductivity to the volumetric specific heat of the material [24]. It measures the rate of heat transfer from an exposed surface of a material to inner layers The larger the diffusivity, the faster the temperature rise at a certain depth in the material [12]. Similar to thermal conductivity and specific heat, thermal diffusivity varies with temperature rise in the material.

Thermal diffusivity, α, can be calculated using the relation

$$\alpha = \frac{k}{\rho c_p},$$

(1)

Where k is thermal conductivity, ρ is density, and c_p is specific heat of the material.

The density, in an oven-dry condition, is the mass of a unit volume of the material, comprising the solid itself and the air-filled pores. With increasing temperature, materials such as concrete that have high amount of moisture will experience loss of mass resulting from evaporation of moisture due to chemical reactions. Assuming that the material is isotropic with respect to its dilatometric behavior, its density (or mass) at any temperature can be calculated from thermogravimetric and dilatometric curves [24].

Mechanical Properties

The mechanical properties that determine the fire performance of RC members are compressive and tensile strength, modulus of elasticity, and stress-strain response of constituent materials at elevated temperatures.

Compressive strength of concrete at an elevated temperature is of primary interest in fire resistance design. Compressive strength of concrete at ambient temperature depends upon water-cement ratio, aggregate-paste interface transition zone, curing conditions, aggregated type and size, admixture types, and type of stress [25]. At high temperature, compressive strength is highly influenced by room temperature strength, rate of heating, and binders in batch mix (such as silica fume, fly ash, and slag). Unlike thermal properties at high temperature, the mechanical properties of concrete are well researched. The strength degradation in HSC is not consistent and there are significant variations in strength loss, as reported by various authors.

The tensile strength of concrete is much lower than compressive strength, due to ease with which cracks can propagate under tensile loads [26]. Concrete is weak in tension, and for NSC, tensile strength is only 10% of its compressive strength and for HSC tensile strength ratio is further reduced. Thus, tensile strength of concrete is often neglected in strength calculations at room and elevated temperatures. However, it is an important property, because cracking in concrete is generally due to tensile stresses and the structural damage of the member in tension is often generated by progression in microcracking [26]. Under fire conditions tensile strength of concrete can be even more crucial in cases where fire induced spalling occurs in a concrete structural member [27]. Tensile strength of concrete is dependent on almost same factors as compressive strength of concrete [28, 29].

Another property that influences fire resistance is the modulus of elasticity of concrete which decreases with temperature. At high temperature, disintegration of hydrated cement products and breakage of bonds in the microstructure of cement paste reduce elastic modulus and the extent of reduction depends on moisture loss, high temperature creep, and type of aggregate.

Deformation Properties

The deformation properties that determine the fire performance of reinforced concrete members are thermal expansion and creep of the concrete and reinforcement at elevated temperatures. In addition, transient strain that occurs at elevated temperatures in concrete can enhance deformations in fire exposed concrete structural members.

Thermal expansion characterizes the expansion (or shrinkage) of a material caused by heating and is defined as the expansion (shrinkage) of unit length of a material when the temperature of concrete is raised by one degree. The coefficient of thermal expansion is defined as the percentage change in length of a specimen per degree temperature rise. The expansion is considered to be positive when the material elongates and is considered negative (shrinkage) when it shortens. In general, the thermal expansion of a material is dependent on the temperature and is evaluated through the dilatometric curve, which is a record of the fractional change of a linear dimension of a solid at a steadily increasing or decreasing temperature [24]. Thermal expansion is an important property to predict thermal stresses that get introduced in a structural member under fire conditions. Thermal expansion of concrete is generally influenced by cement type, water content, aggregate type, temperature, and age [15, 30].

Creep, often referred to as creep strain, is defined as the time-dependent plastic deformation of the material. At normal stresses and ambient temperatures, deformations due to creep are not significant. At higher stress levels and at elevated temperatures, however, the rate of deformation caused by creep can be substantial. Hence, the main factors that influence creep are the temperature, the stress level, and their duration [31]. The creep of concrete is due to the presence of water in its microstructure [32]. There is no satisfactory explanation for the creep of concrete at elevated temperatures.

Transient strain occurs during the first time heating of concrete and is independent of time. It is essentially caused by thermal incompatibilities between the aggregate and the cement paste [6]. Transient strain of concrete, similar to that of high temperature creep, is a complex phenomenon and is influenced by factors such as temperature, strength, moisture content, loading, and mix proportions.

Spalling

In addition to thermal, mechanical, and deformation properties, another property that has a significant influence on the fire performance of a concrete structural member is spalling [33]. This property is unique to concrete and can be a governing factor in determining the fire resistance of an RC structural member [34]. Spalling is defined as the breaking up of layers (pieces) of concrete from the surface of a concrete member when it is exposed to high and rapidly rising temperatures such as those encountered in fires. The spalling can occur soon after exposure to rapid heating and can be accompanied by violent explosions or it may happen during later stages of fire when concrete has become so weak after heating such that, when cracks develop, pieces of concrete fall off from the surface of concrete member. The consequences are limited as long as the extent of damage is small, but extensive spalling may lead to early loss of stability and integrity. Further, spalling exposes deeper layers of concrete to fire temperatures, thereby increasing the rate of transmission of heat to the inner layers of the member, including the reinforcement. When the reinforcement is directly exposed to fire, the temperatures in the reinforcement rise at a very high rate leading to a faster decrease in strength (capacity) of the structural member. The loss of strength in the reinforcement, combined with the loss of concrete due to spalling, significantly decreases the fire resistance of a structural member [35, 36].

While spalling might occur in all concrete types, HSC is more susceptible to fire induced spalling than NSC because of its low permeability and lower water-cement ratio, as compared to NSC. The fire induced spalling is further dependent on a number of factors including permeability of concrete, type of fire exposure, and tensile strength of concrete [34, 37–40]. Thus, information on permeability and tensile strength of concrete, which vary with temperature, are crucial for predicting fire induced spalling in concrete members.

THERMAL PROPERTIES OF CONCRETE AT ELEVATED TEMPERATURES

Thermal properties that govern temperature dependent properties in concrete structures are thermal conductivity, specific heat (or heat capacity), and mass loss. These properties are significantly influenced by the aggregate type, moisture content, and composition of concrete mix. There have been numerous test programs for characterizing thermal properties of concrete at elevated temperatures [16, 41–44]. A detailed review on the effect of temperature on thermal properties of different concrete types is given by Khaliq [45], Kodur et al. [46], and Flynn [47].

Thermal Conductivity

Thermal conductivity of concrete at room temperature is in the range of 1.4 and 3.6 W/m°K and varies with temperature [18]. Figure 1 illustrates the variation of thermal conductivity of NSC as a function of temperature based on published test data and empirical relations. The test data is compiled by Khaliq [45] from different sources based on experimental data [16, 20, 21, 24, 44, and 48] and empirical relations in different standards [4, 15]. The variation in measured test data is depicted through the shaded area in Figure 1 and this variation in reported data on thermal conductivity is mainly attributed to moisture content, type of aggregate, test conditions, and measurement techniques used in experiments [15, 18–20, 41]. It should be noted that there are very few standardized methods available for measuring thermal properties. Also plotted in Figure 1 is both the upper and lower bound values of thermal conductivity as per EC2 provisions and this range is for all aggregate types. However, thermal conductivity shown in Figure 1, as per ASCE relations, is applicable for carbonate aggregates concrete.

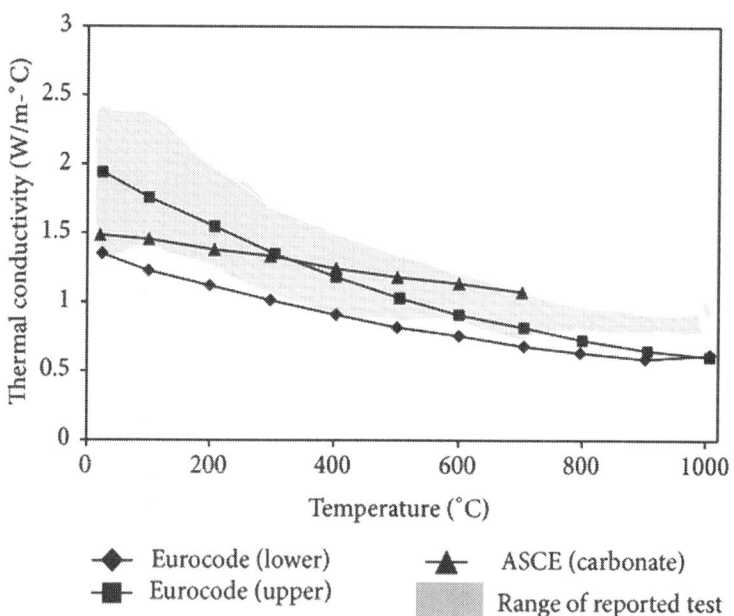

Figure 1: Variation in thermal conductivity of normal strength concrete as a function of temperature.

Overall thermal conductivity decreases gradually with temperature and this decrease is dependent on the concrete mix properties, specifically moisture content and permeability. This decreasing trend in thermal conductivity can be attributed to variation of moisture content with increase in temperature [18].

Thermal conductivity of HSC is higher than that of NSC due to low w/c ratio and use of different binders in HSC [49]. Generally, thermal conductivity of HSC is in the range between 2.4 and 3.6 W/m°K at room temperature. Thermal conductivity for fiber-reinforced concretes (with both steel and polypropylene fibers) almost follow a similar trend as that of plain concrete and lie closer to that of HSC. Therefore, it is deduced that there is no significant effect of fibers on thermal conductivity of concrete in a 20–800°C temperature range [27].

Specific Heat

The specific heat of concrete at room temperature varies in the range of 840 J/kg·K and 1800 J/kg·K for different aggregate types. Often specific heat is expressed in terms of thermal capacity which is the product of specific heat and density of concrete. The specific heat property is sensitive to various physical and chemical transformations that take place in concrete at elevated temperatures. This includes the vaporization of free water at about 100°C, the dissociation of $Ca(OH)_2$ into CaO and H_2O between 400–500°C, and the quartz transformation of some aggregates above 600°C [24]. Specific heat is therefore highly dependent on moisture content and considerably increases with higher water to cement ratio.

Khaliq and Kodur [27] compiled measured specific heat of different concretes from various studies [16, 20, 24, 41, 44, and 48]. Figure 2 illustrates the variation of specific heat for NSC with temperature as reported in various studies based on test data and different standards. The specific heat of concrete type remains almost constant up to 400°C, followed by increases of up to about 700°C and then remains constant between 700 and 800°C range. Of the various factors, aggregate type has a significant influence on the specific heat (thermal capacity) of concrete. This effect is captured in ASCE specified relations for specific heat of concrete [15]. Carbonate aggregate concrete has higher specific heat (heat capacity) in 600–800°C temperature range and this is caused by an endothermic reaction, which results from decomposition of dolomite and absorbs a large amount of energy [12]. This high heat capacity in carbonate aggregate concrete helps to minimize spalling and enhance fire resistance of structural members.

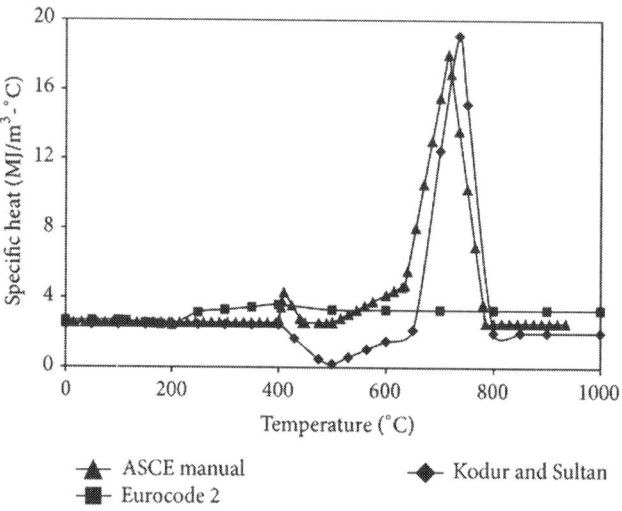

Figure 2: Variation in specific heat of normal strength concrete as a function of temperature.

As compared to NSC, HSC exhibits slightly lower specific heat throughout the 20–800°C temperature range [41]. The presence of fibers also has a minor influence on the specific heat of concrete. For concrete with polypropylene fibers, the burning of polypropylene fibers produces microchannels for release of vapor; and hence the amount of heat absorbed is less for dehydration of chemically bound water; thus its specific heat is reduced in the temperature range of 600–800°C. However, concrete with steel fibers displays a higher specific heat in the 400–800°C temperature range, which can be attributed to additional heat absorbed for dehydration of chemically bound water.

Mass Loss

Depending on the density, concretes are usually subdivided into two major groups: (1) normal-weight concretes with densities in 2150 to 2450 kg·m^{-3} range; and (2) lightweight concretes with densities between 1350 and 1850 kg·m^{-3} The density or mass of concrete decreases with increasing temperature due to loss of moisture. The retention in mass of concrete at elevated temperatures is highly influenced by the type of aggregate [21, 44].

Figure 3 illustrates the variation in mass of concrete as a function of temperature for concretes made with carbonate and siliceous aggregates. The mass loss is minimal for both carbonate and siliceous aggregate concretes up to about 600°C. However, the type of aggregate has significant influence on mass loss in concretes beyond 600°C. In the case of siliceous aggregate concrete, mass loss is insignificant even above 600°C. However, beyond 600°C, carbonate aggregate concrete experiences a larger percentage of mass loss as compared to siliceous aggregate concrete. This higher percentage of mass loss in carbonate aggregate concrete is attributed to dissociation of dolomite in carbonate aggregate at around 600°C [12].

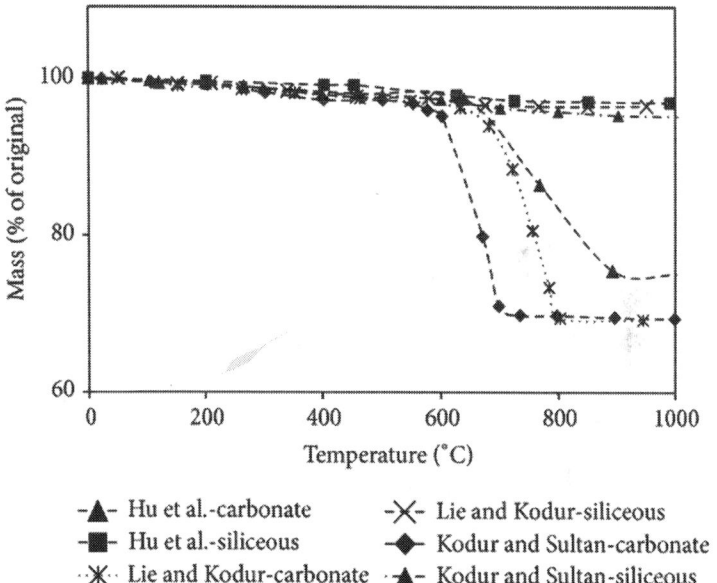

Figure 3: Variation in mass of concrete with different aggregates as a function of temperature.

The strength of concrete does not have a significant influence on mass loss, and hence HSC exhibits a similar trend in mass loss as that of NSC. The mass loss for fiber-reinforced concrete is also similar to plain concrete of up to about 800°C. Above 800°C, the mass loss in steel fiber-reinforced HSC is slightly lower than that of plain HSC.

MECHANICAL PROPERTIES OF CONCRETE AT ELEVATED TEMPERATURES

The mechanical properties that are of primary interest in fire resistance design are compressive strength, tensile strength, elastic modulus, and stress-strain response in compression. Mechanical properties of concrete at elevated temperatures have been studied extensively in the literature in comparison to thermal properties [12, 39, 50–52]. High temperature mechanical property tests are generally carried out on concrete specimens that are typically cylinders or cubes of different sizes. Unlike room temperature property measurements, where there are specified specimen sizes as per standards, the high temperature mechanical properties are usually carried out on a wide range of specimen sizes due to a lack of standardized test specifications for undertaking high temperature mechanical property tests [53, 54].

Compressive Strength

Figures 4 and 5 illustrate the variation of compressive strength ratio for NSC and HSC at elevated temperatures, respectively, with upper and lower bounds (of shaded area) showing range variation in reported test data. Also plotted in these figures is the variation of compressive strength as obtained using Eurocode [4], ASCE [15], and Kodur et al. [46] relations; Figure 4 shows a large but uniform variation of the compiled test data for NSC throughout a 20–800°C temperature range. However, Figure 5 shows a larger variation in compressive strength of HSC with a temperature in the range of 200°C to 500°C and less variation above 500°C. This is mainly because fewer test data points were reported for HSC for temperatures higher than 500°C, either due to the occurrence of spalling in concrete or due to limitations in the test apparatus. However, a wider variation is observed for NSC in this temperature range (above 500°C) when compared to HSC as seen from Figures 4 and 5. This is mainly because of the higher number of test data points reported for NSC in the literature and also due to the lower tendency of NSC to spall under fire. Overall the variation in compressive strength mechanical properties of concrete at high-temperatures is quite high. These variations from different tests can be attributed to using different heating or loading rates, specimen size and curing, condition at testing (moisture content and age of specimen), and the use of admixtures.

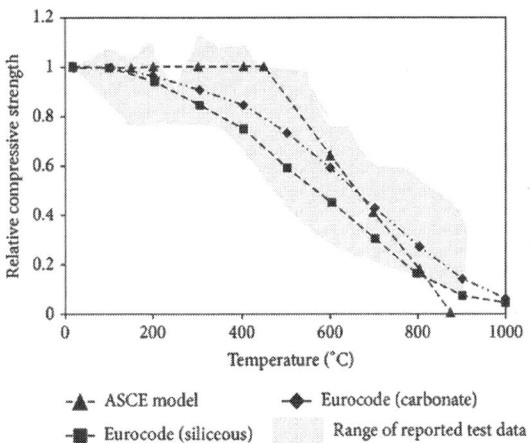

Figure 4: Variation of relative compressive strength of normal strength concrete as a function of temperature.

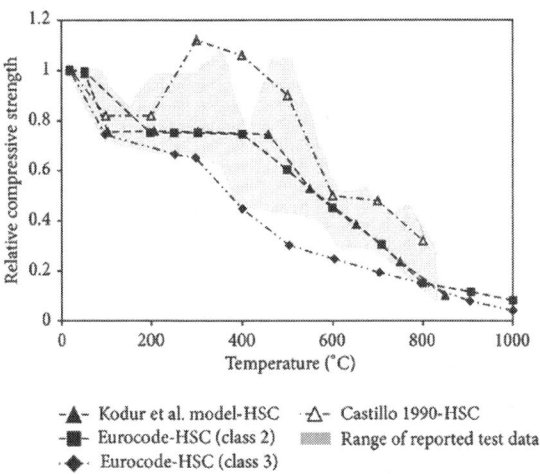

Figure 5: Variation in relative compressive strength of high strength concrete as a function of temperature.

In the case of NSC, the compressive strength of concrete is marginally affected by a temperature of up to 400°C. NSC is usually highly permeable and allows easy diffusion of pore pressure as a result of water vapor. On the other hand, the use of different binders in HSC produces a superior and dense microstructure with less amount of calcium hydroxide which ensures a beneficial effect on compressive strength at room temperature [55]. Binders such as use of slag and silica fume give the best results to improve compressive strength at room temperature which is attributed to a dense microstructure. However, as mentioned earlier, the compact microstructure is highly impermeable and under high temperature becomes detrimental as it does not allow moisture to escape resulting in build-up of pore pressure and rapid development of microcracks in HSC leading to a faster deterioration of strength and occurrence of spalling [27, 56, 57]. The presence of steel fibers in concrete helps to slow down strength loss at elevated temperatures [44, 58].

Among the factors that directly affect compressive strength at elevated temperatures are initial curing, moisture content at the time of testing, and the addition of admixtures and silica fume to the concrete mix [59–63]. These factors are not addressed in the literature and there is no test data that shows the influence of these factors on the high-temperature mechanical properties of concrete.

Another main reasoning for the significant variation in the high-temperature strength properties of concrete is the use of different testing conditions (such as heating rate and strain rate) and test procedures (hot strength test and residual strength test) due to a lack of standardized test methods for carrying out property tests [46].

Tensile Strength

The tensile strength of concrete is much lower than that of compressive strength, and hence tensile strength of concrete is often neglected in strength calculations at room and elevated temperatures. However, from fire resistance point of view, it is an important property, because cracking in concrete is generally due to tensile stresses and the structural damage of the member in tension is often generated by progression in microcracking [26]. Under fire conditions, tensile strength of concrete can be even more crucial in cases where fire induced spalling occurs in concrete member [27]. Thus, information on tensile strength of HSC, which varies with temperature, is crucial for predicting fire induced spalling in HSC members.

Figure 6 illustrates the variation of splitting tensile strength ratio of NSC and HSC as a function of temperature as reported in previous studies and Eurocode provisions [4, 64–66]. The ratio of tensile strength at a given temperature, to that at room temperature, is plotted in Figure 6. The shaded portion in this plot shows a range of variation in splitting tensile strength as obtained by various researchers for NSC with conventional aggregates. The decrease in tensile strength of NSC with temperature can be attributed to weak microstructure of NSC allowing initiation of microcracks. At 300°C, concrete loses about 20% of its initial tensile strength. Above 300°C, the tensile strength of NSC decreases at a rapid rate due to a more pronounced thermal damage in the form of microcracks and reaches to about 20% of its initial strength at 600°C.

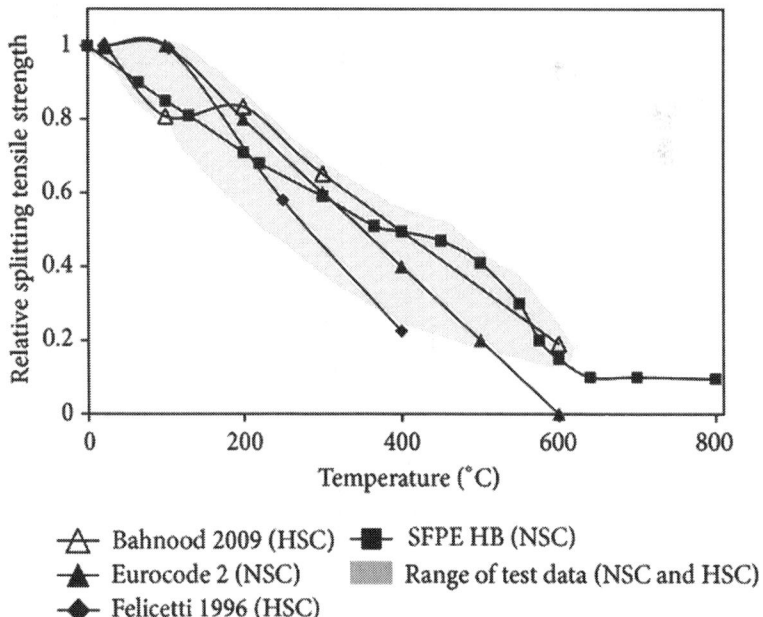

Figure 6: Variation in relative splitting tensile strength of concrete as a function of temperature.

HSC experiences a rapid loss of tensile strength at higher temperatures due to development of pore pressure in dense microstructured HSC [55]. The addition of steel fibers to concrete enhances its tensile strength and the increase can be up to 50% higher at room temperature [67, 68]. Further, the tensile strength of steel fiber-reinforced concrete decreases at a lower rate than that of plain concrete throughout the temperature range of 20–800°C [69]. This increased tensile strength can delay the propagation of cracks in steel fiber-reinforced concrete structural members and is highly beneficial when the member is subjected to bending stresses.

Elastic Modulus

The modulus of elasticity (E) of various concretes at room temperature varies over a wide range, 5.0×10^3 to 35.0×10^3 MPa, and is dependent mainly on the water-cement ratio in the mixture, the age of concrete, the method of conditioning, and the amount and nature of the aggregates. The modulus of elasticity decreases rapidly with the rise of temperature, and the fractional decline does not depend significantly on the type of aggregate [70]. From other surveys [38, 71], it appears, however, that the modulus of elasticity of normal-weight concretes decreases at a higher pace with the rise of temperature than that of lightweight concretes.

Figure 7 illustrates variation of ratio of elastic modulus at target temperature to that at room temperature for NSC and HSC [4, 19, and 72]. It can be seen from the figure that the trend of loss of elastic modulus of both concretes with temperature is similar, but there is a significant variation in the reported test data. The degradation modulus in both NSC and HSC can be attributed to excessive thermal stresses and physical and chemical changes in concrete microstructure.

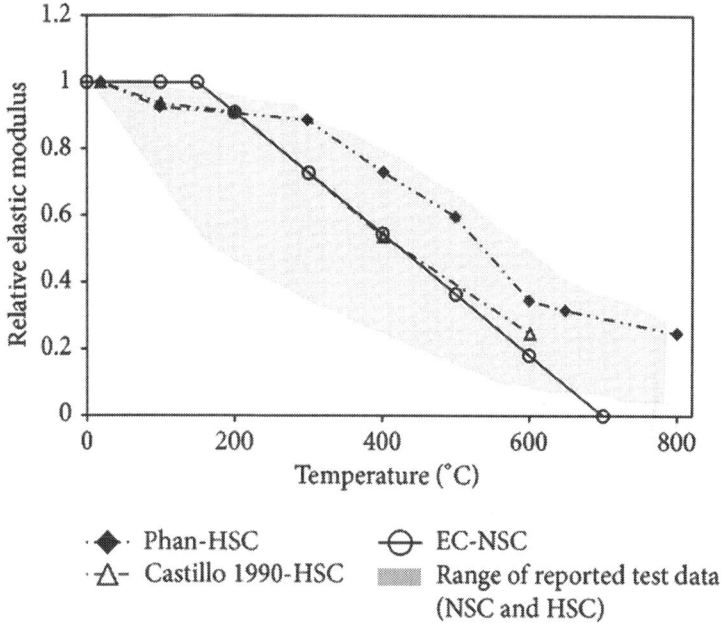

Figure 7: Variation in elastic modulus of concrete as a function of temperature.

Stress-Strain Response

The mechanical response of concrete is usually expressed in the form of stress-strain relations, which are often used as input data in mathematical models for evaluating the fire resistance of concrete structural members. Generally, because of a decrease in compressive strength and increase in ductility of concrete, the slope of stress-strain curve decreases with increasing temperature. The strength of concrete has a significant influence on stress-strain response both at room and elevated temperatures.

Figures 8 and 9 illustrate stress-strain response of NSC and HSC, respectively, at various temperatures [72, 73]. At all temperatures both NSC and HSC exhibit a linear response followed by a parabolic response till peak stress, and then a quick descending portion prior to failure. In general, it is established that HSC has steeper and more linear stress-strain curves in comparison to NSC in 20–800°C. The temperature has a significant effect on the stress-strain response of both NSC and HSC, as with the rate of rise in temperature. The strain corresponding to peak stress starts to increase, especially above 500°C. This increase is significant and the strain at peak stress can reach four times the strain at room temperature. HSC specimens exhibit a brittle response as indicated by postpeak behavior of stress-strain curves shown in Figure 9 [74]. In the case of fiber-reinforced concrete, especially with steel fibers, the stress-strain response is more ductile.

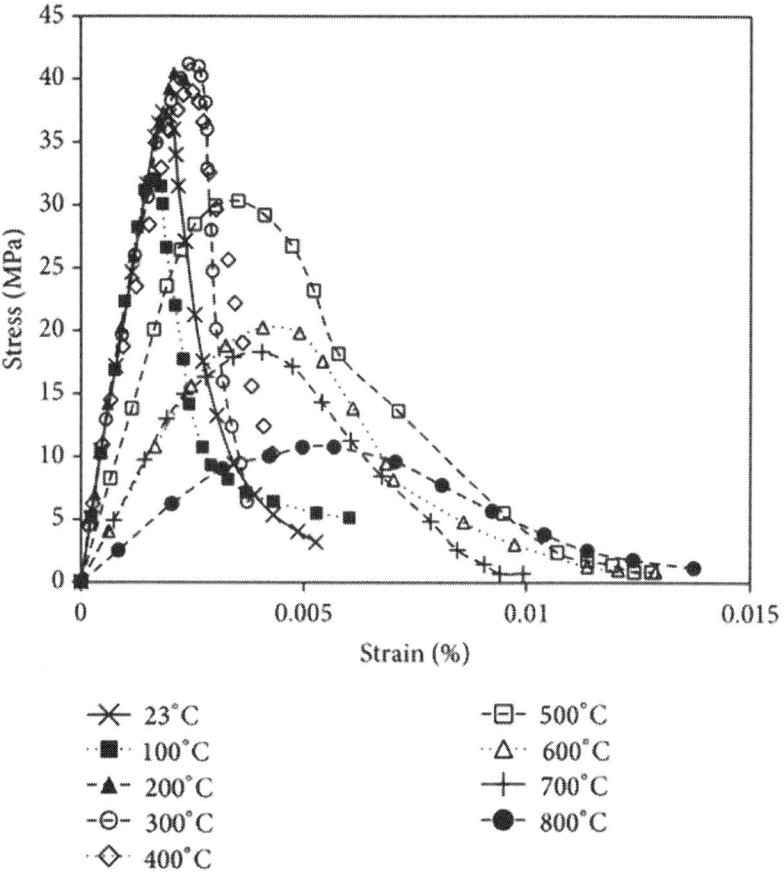

Figure 8: Stress-strain response of normal strength concrete at elevated temperatures.

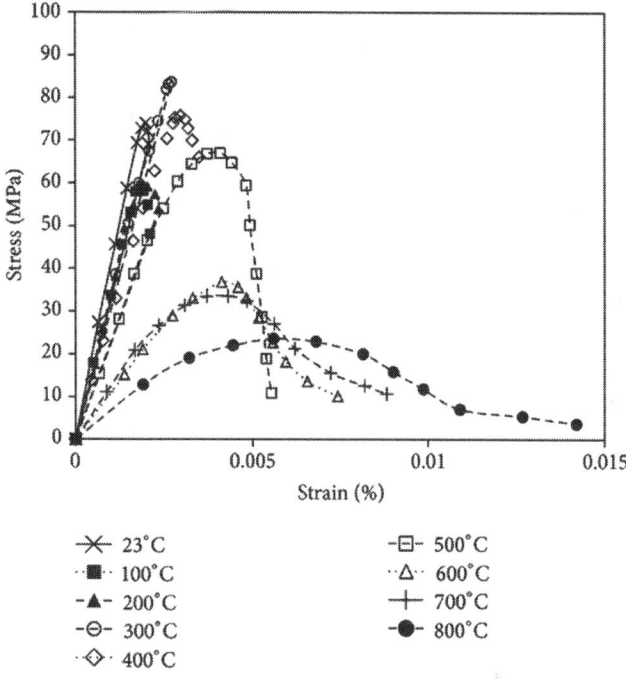

Figure 9: Stress-strain response of high strength concrete at elevated temperatures.

DEFORMATION PROPERTIES OF CONCRETE AT ELEVATED TEMPERATURES

Deformation properties that include thermal expansion, creep strain, and transient strain are highly dependent on the chemical composition, the type of aggregate, and the chemical and physical reactions that occur in concrete during heating [75].

Thermal Expansion

Concrete generally undergoes expansion when subjected to elevated temperatures. Figure 10 illustrates the variation of thermal expansion in NSC with temperature [4, 15], where the shaded portion indicates the range of test data reported by different researchers [46, 76]. The thermal expansion of concrete increases from zero at room temperature to about 1.3% at 700°C and then generally remains constant through 1000°C. This increase is substantial in the 20–700°C temperature range and is mainly due to high thermal expansion resulting from constituent aggregates and cement paste in concrete. Thermal expansion of concrete is complicated by other contributing factors such as additional volume changes caused by variation in moisture content, by chemical reactions (dehydration, change of composition), and by creep and microcracking resulting from nonuniform thermal stresses [18]. In some cases, thermal shrinkage can also result from loss of water due to heating, along with thermal expansion, and this might lead to the overall volume change to be negative, that is, shrinkage rather than expansion.

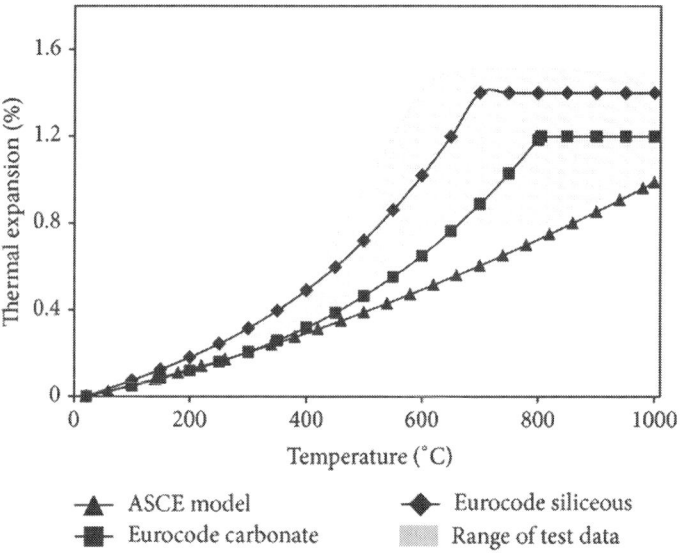

Figure 10: Variation in linear thermal expansion of normal strength concrete as a function of temperature.

Eurocodes [4] accounts for the effect of type of aggregate on variation of thermal expansion than of concrete with temperature. Concrete made with siliceous aggregate has a higher thermal expansion than that of carbonate aggregate concrete. However, ASCE provisions [15] provide only one variation for both siliceous and carbonate aggregate concrete.

The strength of concrete and presence of fiber have moderate influence on thermal expansion. The rate of expansion for HSC and fiber-reinforced concrete slows down between 600–800°C; however the rate of thermal expansion increases again above 800°C. The slowdown in thermal expansion in the 600–800°C range is attributed to the loss of chemically bound water in hydrates, and the increase in expansion above 800°C is attributed to a softening of concrete and excessive micro- and macrocrack development [77].

Creep and Transient Strains

Time-dependent deformations in concrete such as creep and transient strains get highly enhanced at elevated temperatures under compressive stresses [18]. Creep in concrete under high temperatures increases due to moisture movement out of concrete matrix. This phenomenon is further intensified by moisture dispersion and loss of bond in cement gel (C–S–H). Therefore, the process of creep is caused and accelerated mainly by two processes: (1) moisture movement and dehydration of concrete due to high temperatures and (2) acceleration in the process of breakage of bond.

Transient strain occurs during the first time heating of concrete, but it does not occur upon repeated heating [78]. Exposure of concrete to high temperature induces complex changes in the moisture content and chemical composition of the cement paste. Moreover, there exists a mismatch in the thermal expansion between the cement paste and the aggregate. Therefore, factors such as changes in chemical composition of concrete and mismatches in thermal expansion lead to internal stresses and microcracking in the concrete constituents (aggregate and cement paste) and results in transient strain in the concrete [75].

A review of the literature shows that there is limited information on creep and transient strain of concrete at elevated temperatures [46]. Some data on the creep of concrete at elevated temperatures is available from the work of Cruz, [70], MareÂchal [79], Gross [80], and Schneider et al. [81]. Anderberg and Thelandersson [82] carried out tests to evaluate transient and creep strains under elevated temperatures. They found that predried specimens at 45 and 67.5%

of load stress level were less liable to deformation in the "positive direction" (expansion) under load. At 22.5% preload, specimens displayed no significant difference of strains. They also found that the influence of water saturation was not very significant except for free thermal expansion (0% preload), which was found to be smaller for water saturated specimens.

Khoury et al. [78] studied creep strain of initially moist concretes at four load levels measured during first heating at 1°C/min. An important feature of these results was that a considerable contraction under load was observed as compared to free (unloaded) thermal strains. This contraction is referred to as the "load-induced thermal strain" and the actual thermal strain is considered to consist of total thermal strain minus the load-induced thermal strain.

Schneider [75] also investigated the effect of transient and creep restraint on deformation of concrete. He concluded that the transient test for measuring total deformation or restraint of concrete has the strongest relation to building fires and is supposed to give the most realistic data with direct relevance to fire. The important conclusions from the study are that (1) water to cement ratio and original strength are of little importance on creep deformations under transient conditions, (2) aggregate to cement ratio has a great influence on the strains and critical temperatures: the harder the aggregate the lower the thermal expansion; therefore total deformation in transient state will be lower; and (3) curing conditions are of great importance in the 20–300°C range: air-cured and oven-dried specimens have lower transient and creep strains than water cured specimens.

Anderberg and Thelandersson [82] developed constitutive models for creep and transient strains in concrete at elevated temperatures. These equations for creep and transient strain at elevated temperatures as suggested by Anderberg and Thelandersson [82] are

$$\varepsilon_{cr} = \beta_1 \frac{\sigma}{f_{c,T}} \sqrt{t} e^{d(T-293)},$$

$$\varepsilon_{tr} = k_2 \frac{\sigma}{f_{c,20}} \varepsilon_{th},$$

(2)

Where ε_{cr} = creep strain, ε_{tr} = transient strain, $\beta_1 = 6.28 \times 10^{-6}$ s$^{-0.5}$, $d = 2.658 \times 10^{-3}$ K^{-1}, T = concrete temperature (°K) at time t (s), f_c =

concrete strength at temperature T, σ = stress in the concrete at the current temperature, k_2 = a constant ranges between 1.8 and 2.35, ε_{th} = thermal strain, and $f_{c,20}$ = concrete strength at room temperature.

The above discussed information on high temperature creep and transient strain is mostly developed for NSC. There is still a lack of test data and models on the effect of temperature on creep and transient strain in HSC and fiberreinforced concrete.

FIRE INDUCED SPALLING

A review of the literature presents a conflicting picture on the occurrence of fire induced spalling and also on the exact mechanism of spalling in concrete. While some researchers reported explosive spalling in concrete structural members exposed to fire, a number of other studies reported little or no significant spalling. One possible explanation for this confusing trend of observations is the large number of factors that influence spalling and their interdependency. However, most researchers agree that major causes for fire induced spalling in concrete are low permeability of concrete and moisture migration in concrete at elevated temperatures.

There are two broad theories by which the spalling phenomenon can be explained [83].

Pressure Build-Up: Spalling is believed to be caused by the build-up of pore pressure during heating [83–85]. The extremely high water vapor pressure, generated during exposure to fire, cannot escape due to the high density and compactness (and low permeability) of higher strength concrete When the effective pore pressure (porosity times pore pressure) exceeds the tensile strength of concrete, chunks of concrete fall off from the structural member. This pore pressure is considered to drive progressive failure; that is, the lower the permeability of concrete, the greater the fire induced spalling. This falling-off of concrete chunks can often be explosive depending on fire and concrete characteristics [38, 86].

Restrained Thermal Dilatation: This hypothesis considers that spalling results from restrained thermal dilatation close to the heated surface, which leads to the development of compressive stresses parallel to the heated surface. These compressive stresses are released by brittle fracture of concrete (spalling). The pore pressure can play a significant role on the onset of instability in the form of explosive thermal spalling [87].

Although spalling might occur in all concretes, high-strength concrete is believed to be more susceptible to spalling than normal-strength concrete because of its low permeability and low water-cement ratio [88, 89]. The high water vapor pressure, generated due to a rapid rise in temperature, cannot escape due to high density (and low permeability) of HSC, and this pressure build-up often reaches the saturation vapor pressure. At 300°C, the pore pressure can reach up to 8 MPa; such internal pressures are often too high to be resisted by the HSC mix having a tensile strength of approximately 5 MPa [84]. The drained conditions at the heated surface and the low permeability of concrete lead to strong pressure gradients close to the surface in the form of the so-called "moisture clog" [38, 86]. When the vapor pressure exceeds the tensile strength of concrete, chunks of concrete fall off from the structural member. In a number of test observations on HSC columns, it has been found that spalling is often of an explosive nature [19, 90]. Hence, spalling is one of the major concerns in the use of HSC in building applications and should be properly accounted for in evaluating fire performance [91]. Spalling in NSC and HSC columns is compared in Figure 11 using data obtained from full-scale fire tests on loaded columns [92]. It can be seen that the spalling is quite significant in fire exposed HSC column.

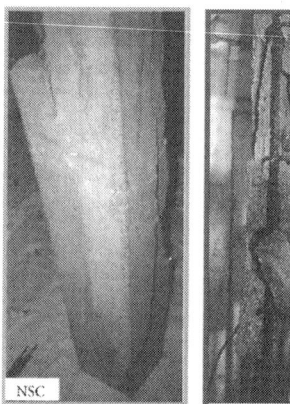

Figure 11: Relative spalling in NSC and HSC columns under fire conditions.

The extent of spalling depends on a number of factors including strength, porosity, density, load level, fire intensity, aggregate type, relative humidity, amount of silica fume, and other admixtures [34, 93, and 94]. Many of these factors are interdependent and this makes prediction of spalling quite complex. The variation of porosity with temperature is the most important property needed for predicting spalling performance of HSC [33]. Noumowé et al. carried out porosity measurements on NSC and HSC specimens, using a mercury porosimeter, at various temperatures [88, 95].

Based on limited fire tests, researchers have suggested that spalling in HSC can be minimized by adding polypropylene fibres to the HSC mix [85, 96–101]. The polypropylene fibers melt when temperatures in concrete reach about 160–170°C and this creates pores in concrete that are sufficient for relieving vapor pressure developed in the concrete. Another alternative for limiting fire induced spalling in HSC columns is through the use of bent ties, where ties are bent at 135° into the concrete core [102].

RELATIONS OF HIGH TEMPERATURE PROPERTIES OF CONCRETE

There are limited constitutive relations for high-temperature properties of concrete in codes and standards that can be used for fire design. These relations can be found in the ASCE manual [15] and Eurocode 2 [4]. Kodur et al. [46] have compiled different relations that are available for thermal, mechanical, and deformation of concrete at elevated temperatures.

There are some differences in the constitutive relationships for high-temperature properties of concrete used in European and American standards. The constitutive relations in Eurocode are applicable for NSC and HSC, while the relations in the ASCE manual of practice are for NSC only. The constitutive relationships for high-temperature properties of concrete specified in the Eurocode and the ASCE manual are summarized in Table1. In addition to these constitutive models, Kodur et al. [93] proposed constitutive relations for HSC, which are an extension to ASCE relations for NSC. These relations for HSC are also included in Table 1.

Handbook of Materials Structures, Properties, Processing and...

Table 1: Constitutive relationships of high-temperature properties of concrete

	NSC—ASCE Manual 1992	HSC—Kodur et al. 2004 [10]	NSC and HSC—EN1992-1-2: 2004 [4]
Stress-strain relationships	$\sigma_c = \begin{cases} f'_{c,T}\left[1-\left(\dfrac{\varepsilon-\varepsilon_{max,T}}{\varepsilon_{max,T}}\right)^2\right], & \varepsilon \le \varepsilon_{max,T} \\ f'_{c,T}\left[1-\left(\dfrac{\varepsilon_{max,T}-\varepsilon}{3\varepsilon_{max,T}}\right)^2\right], & \varepsilon > \varepsilon_{max,T} \end{cases}$ $f'_{c,T} = \begin{cases} f'_c, & 20°C \le T \le 450°C \\ f'_c\left[2.011-2.353\left(\dfrac{T-20}{1000}\right)\right], & 450°C < T \le 874°C \\ 0, & 874°C < T \end{cases}$ $\varepsilon_{max,T} = 0.0025 + (6.0T + 0.04T^2)\times 10^{-6}$	$\sigma_c = \begin{cases} f'_{c,T}\left[1-\left(\dfrac{\varepsilon_{max,T}-\varepsilon}{\varepsilon_{max,T}}\right)^H\right], & \varepsilon \le \varepsilon_{max,T} \\ f'_{c,T}\left[1-\left(\dfrac{30(\varepsilon-\varepsilon_{max,T})}{(130-f'_c)\varepsilon_{max,T}}\right)^2\right], & \varepsilon > \varepsilon_{max,T} \end{cases}$ $f'_{c,T} = \begin{cases} f'_c[1.0-0.003125(T-20)], & T < 100°C \\ 0.75f'_c, & 100°C \le T \le 400°C \\ f'_c[1.33-0.00145T], & 400°C < T \end{cases}$ $\varepsilon_{max,T} = 0.0018 + (6.7f'_c + 6.0T + 0.03T^2)\times 10^{-6}$ $H = 2.28 - 0.012f'_c$	$\sigma_c = \dfrac{3\varepsilon f'_{c,T}}{\varepsilon_{c1,T}\left(2+(\varepsilon/\varepsilon_{c1,T})^3\right)}, \; \varepsilon \le \varepsilon_{c1,T}$ For $\varepsilon_{c1(T)} \le \varepsilon \le \varepsilon_{cu1(T)}$, the Eurocode permits the use of linear as well as nonlinear descending branches in the numerical analysis. For the parameters in this equation refer to Table 2
Thermal capacity	Siliceous aggregate concrete: $\rho c = \begin{cases} 0.005T+1.7, & 20°C \le T \le 200°C \\ 2.7, & 200°C < T \le 400°C \\ 0.013T-2.5, & 400°C < T \le 500°C \\ 10.5-0.013T, & 500°C < T \le 600°C \\ 2.7, & 600°C < T \end{cases}$ Carbonate aggregate concrete: $\rho c = \begin{cases} 2.566, & 20°C \le T \le 400°C \\ 0.1765T-68.034, & 400°C < T \le 410°C \\ 25.00671-0.05043T, & 410°C < T \le 445°C \\ 2.566, & 445°C < T \le 500°C \\ 0.01603T-5.44881, & 500°C < T \le 635°C \\ 0.16635T-100.90225, & 635°C < T \le 715°C \\ 176.07343-0.22103T, & 715°C < T \le 785°C \\ 2.566, & 785°C < T \end{cases}$	Siliceous aggregate concrete: $\rho c = \begin{cases} 0.005T+1.7, & 20°C \le T \le 200°C \\ 2.7, & 200°C < T \le 400°C \\ 0.013T-2.5, & 400°C < T \le 500°C \\ -0.013T+10.5, & 500°C < T \le 600°C \\ 2.7, & 600°C < T \le 635°C \end{cases}$ Carbonate aggregate concrete: $\rho c = \begin{cases} 2.45, & 20°C \le T \le 400°C \\ 0.026T-12.85, & 400°C < T \le 475°C \\ 0.0143T-6.295, & 475°C < T \le 650°C \\ 0.3894T-120.11, & 650°C < T \le 735°C \\ -0.263T+212.4, & 735°C < T \le 800°C \\ 2, & 800°C < T \le 1000°C \end{cases}$	Specific heat (J/kg·C): $c = 900$, for $20°C \le T \le 100°C$, $c = 900+(T-100)$, for $100°C < T \le 200°C$, $c = 1000+(T-200)/2$, for $200°C < T \le 400°C$, $c = 1100$, for $400°C < T \le 1200°C$ Density change (kg/m³): $\rho = \rho(20°C)$ = Reference density for $20°C \le T \le 115°C$ $\rho = \rho(20°C)(1-0.02(T-115)/85)$ for $115°C < T \le 200°C$, $\rho = \rho(20°C)(0.98-0.03(T-200)/200)$ for $200°C < T \le 400°C$, $\rho = \rho(20°C)(0.95-0.07(T-400)/800)$ for $400°C < T \le 1200°C$. Thermal capacity = $\rho \times c$.
Thermal conductivity	Siliceous aggregate concrete: $k_c = \begin{cases} -0.000625T+1.5, & 20°C \le T \le 800°C \\ 1.0, & 800°C < T \end{cases}$ Carbonate aggregate concrete: $k_c = \begin{cases} 1.355, & 20°C \le T \le 293°C \\ -0.001241T+1.7162, & 293°C < T \end{cases}$	Siliceous aggregate concrete: $k_c = 0.85(2-0.0011\,T), \; 20°C \le T \le 1000°C$ Carbonate aggregate concrete: $k_c = \begin{cases} 0.85(2-0.0013T), & 20°C \le T \le 300°C \\ 0.85(2.21-0.002T), & 300°C < T \end{cases}$	All types: Upper limit: $k_c = 2-0.2451(T/100)+0.0107(T/100)^2$, for $20°C \le T \le 1200°C$. Lower limit: $k_c = 1.36-0.136(T/100)+0.0057(T/100)^2$, for $20°C \le T \le 1200°C$.
Thermal strain	All types: $\varepsilon_{th} = [0.004(T^2-400)+6(T-20)]\times 10^{-6}$	All types: $\varepsilon_{th} = [0.004(T^2-400)+6(T-20)]\times 10^{-6}$	Siliceous aggregates: $\varepsilon_{th} = -1.8\times 10^{-4}+9\times 10^{-6}T+2.3\times 10^{-11}T^3$, for $20°C \le T \le 700°C$ $\varepsilon_{th} = 14\times 10^{-3}$, for $700°C < T \le 1200°C$. Calcareous aggregates: $\varepsilon_{th} = -1.2\times 10^{-4}+6\times 10^{-6}T+1.4\times 10^{-11}T^3$, for $20°C \le T \le 805°C$ $\varepsilon_{th} = 12\times 10^{-3}$, for $805°C < T \le 1200°C$.

Table 2: Values of the main parameters of the stress-strain relationships of NSC and HSC at elevated temperatures as specified in EN1992-1-2: 2004 [4]

Temp. °F	Temp. °C	NSC Siliceous agg.			NSC Calcareous agg.			HSC $f'_{c,T}/f'_c(20^0C)$		
		$f'_{c,T}/f'_c(20^0C)$	$\varepsilon_{c1,T}$	$\varepsilon_{cu1,T}$	$f'_{c,T}/f'_c(20^0C)$	$\varepsilon_{c1,T}$	$\varepsilon_{cu1,T}$	Class 1	Class 2	Class 3
68	20	1	0.0025	0.02	1	0.0025	0.02	1	1	1
212	100	1	0.004	0.0225	1	0.004	0.023	0.9	0.75	0.75
392	200	0.95	0.0055	0.025	0.97	0.0055	0.025	0.9	0.75	0.70
572	300	0.85	0.007	0.0275	0.91	0.007	0.028	0.85	0.75	0.65
752	400	0.75	0.01	0.03	0.85	0.01	0.03	0.75	0.75	0.45
932	500	0.6	0.015	0.0325	0.74	0.015	0.033	0.60	0.60	0.30
1112	600	0.45	0.025	0.035	0.6	0.025	0.035	0.45	0.45	0.25
1292	700	0.3	0.025	0.0375	0.43	0.025	0.038	0.30	0.30	0.20
1472	800	0.15	0.025	0.04	0.27	0.025	0.04	0.15	0.15	0.15
1652	900	0.08	0.025	0.0425	0.15	0.025	0.043	0.08	0.113	0.08
1832	1000	0.04	0.025	0.045	0.06	0.025	0.045	0.04	0.075	0.04
2012	1100	0.01	0.025	0.0475	0.02	0.025	0.048	0.01	0.038	0.01
2192	1200	0	—	—	0	—	—	0	0	0

The Eurocode classifies HSC into three classes*, depending on its compressive strength, namely,

- Class 1 for concrete with compressive strength between C55/67 and C60/75,
- Class 2 for concrete with compressive strength between C70/85 and C80/95,
- Class 3 for concrete with compressive strength higher than C90/105.

The strength notation of C55/67 refers to a concrete grade with a characteristic cylinder and cube strength of 55 N/mm^2 and 67 N/mm^2, respectively.

*Note: where the actual characteristic strength of concrete is likely to be of a higher class than that specified in the design; the relative reduction in strength for the higher class should be used for fire design.

A major difference between the European and the ASCE high-temperature constituent relations for concrete is the effect of aggregate type on concrete properties. The Eurocode does not specifically account for the effect of aggregate type on the thermal capacity of concrete at high-temperatures. In the Eurocode, properties such as specific heat, density changes, and hence, heat capacity are considered to be the same for all aggregate types used in concrete. For the thermal conductivity of concrete, the Eurocode proposes upper and lower bound limits without indicating which limit to use for a given aggregate type in concrete. Furthermore, Eurocode classifies HSC into three classes, depending on its compressive strength, namely,

- Class 1 for concrete with compressive strength between C55/67 and C60/75,
- Class 2 for concrete with compressive strength between C70/85 and C80/95,
- Class 3 for concrete with compressive strength higher than C90/105.

SUMMARY

Concrete, at elevated temperatures, undergoes significant physicochemical changes. These changes cause properties to deteriorate at elevated temperatures and introduce additional complexities, such as spalling in HSC. Thus, thermal, mechanical, and deformation properties of concrete change substantially within the temperature range associated with building fires. Furthermore, many of these properties are temperature dependent and sensitive to testing (method) parameters such as heating rate, strain rate, temperature gradient, and so on.

Based on information presented in this chapter, it is evident that high temperature properties of concrete are crucial for modeling fire response of reinforced concrete structures. A good amount of data exists on high temperature thermal, mechanical, and deformation properties of NSC and HSC. However, there is very limited property data on high temperature properties of new types of concrete such as self-consolidated concrete and fly ash concrete at elevated temperatures.

The review on material properties provided in this chapter is a broad outline of currently available information. Additional details related to specific conditions on which these properties are developed can be found in cited references. Also, when using the material properties presented in this chapter, due consideration should be given to batch mix properties and other characteristics, such as heating rate and loading level, because the properties at elevated temperatures depend on a number of factors.

DISCLAIMER

Certain commercial products are identified in this paper in order to adequately specify the experimental procedure. In no case does such identification imply recommendations or endorsement by the author, nor does it imply that the product or material identified is the best available for the purpose.

REFERENCES

1. ACI 216.1, "Code requirements for determining fire resistance of concrete and masonry construction assemblies," ACI 216.1-07/TMS-0216-07, American Concrete Institute, Farmington Hills, Mich, USA, 2007.

2. ACI-318, Building Code Requirements For ReinForced Concrete and Commentary, American Concrete Institute,, Farmington Hills, Mich, USA, 2008.
3. "EN 1991-1-2: actions on structures. Part 1-2: general actions—actions on structures exposed to fire,"Eurocode 1, European Committee for Standardization, Brussels, Belgium, 2002.
4. "EN, 1992-1-2: design of concrete structures. Part 1-2: general rules—structural fire design," Eurocode 2, European Committee for Standardization, Brussels, Belgium, 2004.
5. A. H. Buchanan, Structural Design for Fire Safety, John Wiley and Sons, Chichester, UK, 2002.
6. J. A. Purkiss, Fire Safety Engineering Design of Structures, Butterworth-Heinemann, Elsevier, Oxford, UK, 2007.
7. V. R. Kodur and N. Raut, "Performance of concrete structures under fire hazard: emerging trends," The Indian Concrete Journal, vol. 84, no. 2, pp. 23–31, 2010.
8. "Standard test methods for fire tests of building construction and materials," ASTM E119-08b, ASTM International, West Conshohocken, Pa, USA, 2008.
9. "Fire design of concrete structures—materials, structures and modelling," FIB Bulletin 38, The International Federation for Structural Concrete, Lausanne, Switzerland, 2007.
10. V. K. R. Kodur, T. C. Wang, and F. P. Cheng, "Predicting the fire resistance behaviour of high strength concrete columns," Cement and Concrete Composites, vol. 26, no. 2, pp. 141–153, 2004.
11. V. Kodur, M. Dwaikat, and N. Raut, "Macroscopic FE model for tracing the fire response of reinforced concrete structures," Engineering Structures, vol. 31, no. 10, pp. 2368–2379, 2009.
12. V. R. Kodur and T. Z. Harmathy, "Properties of building materials," in SFPE Handbook of Fire Protection Engineering, P. J. DiNenno, Ed., National Fire Protection Association, Quincy, Mass, USA, 2008.
13. M. B. Dwaikat and V. K. R. Kodur, "Fire induced spalling in high strength concrete beams," Fire Technology, vol. 46, no. 1, pp. 251–274, 2010.
14. "Standard test method for evaluating the resistance to thermal transmission of materials by the guarded heat flow

meter technique," ASTM E1530, ASTM International, West Conshohocken, Pa, USA, 2011.

15. ASCE, Structural Fire Protection, ASCE Committee on Fire Protection, Structural Division, American Society of Civil Engineers, New York, NY, USA, 1992.
16. K.-Y. Shin, S.-B. Kim, J.-H. Kim, M. Chung, and P.-S. Jung, "Thermo-physical properties and transient heat transfer of concrete at elevated temperatures," Nuclear Engineering and Design, vol. 212, no. 1–3, pp. 233–241, 2002.
17. B. Adl-Zarrabi, L. Boström, and U. Wickström, "Using the TPS method for determining the thermal properties of concrete and wood at elevated temperature," Fire and Materials, vol. 30, no. 5, pp. 359–369, 2006.
18. Z. P. Bažant and M. F. Kaplan, Concrete at High Temperatures: Material Properties and Mathematical Models, Longman Group Limited, Essex, UK, 1996.
19. L. T. Phan, "Fire performance of high-strength concrete: a report of the state-of-the-art," Tech. Rep., National Institute of Standards and Technology, Gaithersburg, Md, USA, 1996.
20. T. Z. Harmathy and L. W. Allen, "Thermal properties of selected masonry unit concretes," Journal American Concrete Institution, vol. 70, no. 2, pp. 132–142, 1973.
21. V. R. Kodur and M. A. Sultan, "Thermal propeties of high strength concrete at elevated temperatures,"American Concrete Institute, Special Publication, SP-179, pp. 467–480, 1998.
22. "Standard test method for determining specific heat capacity by differential scanning calorimetry,"ASTM C1269, ASTM International, West Conshohocken, Pa, USA, 2011.
23. ISO/DIS22007-2:2008, "Determination of thermal conductivity and thermal diffusivity, Part 2: transient plane heat source (hot disc) method," ISO, Geneva, Switzerland, 2008.
24. T. Z. Harmathy, "Thermal properties of concrete at elevated temperatures," ASTM Journal of Materials, vol. 5, no. 1, pp. 47–74, 1970.
25. P. K. Mehta and P. J. M. Monteiro, Concrete: Microstructure, Properties, and Materials, McGraw-Hill, New York, NY, USA, 2006.

26. S. Mindess, J. F. Young, and D. Darwin, Concrete, Pearson Education, Upper Saddle River, NJ, USA, 2003.
27. W. Khaliq and V. Kodur, "High temperature mechanical properties of high strength fly ash concrete with and without fibers," ACI Materials Journal, vol. 109, no. 6, pp. 665–674, 2012.
28. A. M. Neville, Properties of Concrete, Pearson Education, Essex, UK, 2004.
29. S. P. Shah, "Do fibers increase the tensile strength of cement-based matrixes?" ACI Materials Journal, vol. 88, no. 6, pp. 595–602, 1991.
30. Z. P. Bazant and J.-C. Chern, "Stress-induced thermal and shrinkage strains in concrete," Journal of Engineering Mechanics, vol. 113, no. 10, pp. 1493–1511, 1987.
31. T. Z. Harmathy, "A comprehensive creep model," Journal of Basic Engineering, vol. 89, no. 3, pp. 496–502, 1967.
32. F. H. Wittmann, Ed., Fundamental Research on Creep and Shrinkage of Concrete, Martinus Nijhoff, The Hague, Netherlands, 1982.
33. M. B. Dwaikat and V. K. R. Kodur, "Hydrothermal model for predicting fire-induced spalling in concrete structural systems," Fire Safety Journal, vol. 44, no. 3, pp. 425–434, 2009.
34. V. K. R. Kodur and L. Phan, "Critical factors governing the fire performance of high strength concrete systems," Fire Safety Journal, vol. 42, no. 6-7, pp. 482–488, 2007.
35. X. Yu, X. Zha, and Z. Huang, "The influence of spalling on the fire resistance of RC structures," Advanced Materials Research, vol. 255–260, pp. 519–523, 2011.
36. V. K. R. Kodur and M. Dwaikat, "Effect of fire induced spalling on the response of reinforced concrete beams," International Journal of Concrete Structures and Materials, vol. 2, no. 2, pp. 71–82, 2008.
37. T. Z. Harmathy, "Moisture and heat transport with particular reference to concrete," NRCC 12143, National Council of Canada, 1971.
38. T. Z. Harmathy, Fire Safety Design and Concrete, John Wiley & Sons, New York, NY, USA, 1993.

39. G. A. Khoury, "Concrete spalling assessment methodologies and polypropylene fibre toxicity analysis in tunnel fires," Structural Concrete, vol. 9, no. 1, pp. 11–18, 2008.
40. P. Kalifa, G. Chéné, and C. Gallé, "High-temperature behaviour of HPC with polypropylene fibres—from spalling to microstructure," Cement and Concrete Research, vol. 31, no. 10, pp. 1487–1499, 2001.
41. V. K. R. Kodur and M. A. Sultan, "Effect of temperature on thermal properties of high-strength concrete," Journal of Materials in Civil Engineering, vol. 15, no. 2, pp. 101–107, 2003.
42. M. G. Van Geem, J. Gajda, and K. Dombrowski, "Thermal properties of commercially available high-strength concretes," Cement, Concrete and Aggregates, vol. 19, no. 1, pp. 38–54, 1997.
43. T. T. Lie and V. R. Kodur, "Thermal properties of fibre-reinforced concrete at elevated temperatures," IR683, IRC, National Research Council of Canada, Ottawa, Canada, 1995.
44. T. T. Lie and V. K. R. Kodur, "Thermal and mechanical properties of steel-fibre-reinforced concrete at elevated temperatures," Canadian Journal of Civil Engineering, vol. 23, no. 2, pp. 511–517, 1996.
45. W. Khaliq, Performance characterization of high performance concretes under fire conditions [Ph.D. thesis], Michigan State University, 2012.
46. V. K. R. Kodur, M. M. S. Dwaikat, and M. B. Dwaikat, "High-temperature properties of concrete for fire resistance modeling of structures," ACI Materials Journal, vol. 105, no. 5, pp. 517–527, 2008.
47. D. R. Flynn, "Response of high performance concrete to fire conditions: review of thermal property data and measurement techniques," Tech. Rep., National Institute of Standards and Technology, Millwood, Va, USA, 1999.
48. T. Harada, J. Takeda, S. Yamane, and F. Furumura, "Strength, elasticity and thermal properties of concrete subjected to elevated temperatures," ACI Concrete For Nuclear Reactor SP, vol. 34, no. 2, pp. 377–406, and 1972.

49. V. Kodur and W. Khaliq, "Effect of temperature on thermal properties of different types of high-strength concrete," Journal of Materials in Civil Engineering, ASCE, vol. 23, no. 6, pp. 793–801, 2011.
50. W. C. Tang and T. Y. Lo, "Mechanical and fracture properties of normal-and high-strength concretes with fly ash after exposure to high temperatures," Magazine of Concrete Research, vol. 61, no. 5, pp. 323–330, 2009.
51. A. Noumowe, "Mechanical properties and microstructure of high strength concrete containing polypropylene fibres exposed to temperatures up to 200°C," Cement and Concrete Research, vol. 35, no. 11, pp. 2192–2198, 2005.
52. M. Li, C. Qian, and W. Sun, "Mechanical properties of high-strength concrete after fire," Cement and Concrete Research, vol. 34, no. 6, pp. 1001–1005, 2004.
53. RILEM TC 129-MHT, "Test methods for mechanical properties of concrete at high temperatures—compressive strength for service and accident conditions," Materials and Structures, vol. 28, no. 3, pp. 410–414, 1995.
54. RILEM TC 129-MHT, "Test methods for mechanical properties of concrete at high temperatures, Part 4—tensile strength for service and accident conditions," Materials and Structures, vol. 33, pp. 219–223, 2000.
55. Y. N. Chan, G. F. Peng, and M. Anson, "Residual strength and pore structure of high-strength concrete and normal strength concrete after exposure to high temperatures," Cement and Concrete Composites, vol. 21, no. 1, pp. 23–27, 1999.
56. C.-S. Poon, S. Azhar, M. Anson, and Y.-L. Wong, "Comparison of the strength and durability performance of normal- and high-strength pozzolanic concretes at elevated temperatures," Cement and Concrete Research, vol. 31, no. 9, pp. 1291–1300, 2001.
57. B. Chen and J. Liu, "Residual strength of hybrid-fiber-reinforced high-strength concrete after exposure to high temperatures," Cement and Concrete Research, vol. 34, no. 6, pp. 1065–1069, 2004.
58. A. Lau and M. Anson, "Effect of high temperatures on high performance steel fibre reinforced concrete,"Cement and Concrete Research, vol. 36, no. 9, pp. 1698–1707, 2006.

59. W. P. S. Dias, G. A. Khoury, and P. J. E. Sullivan, "Mechanical properties of hardened cement paste exposed to temperatures up to 700°C," ACI Materials Journal, vol. 87, no. 2, pp. 160–166, 1990.
60. F. Furumura, T. Abe, and Y. Shinohara, "Mechanical properties of high strength concrete at high temperatures," in Proceedings of the 4th Weimar Workshop on High Performance Concrete, Material Properties and Design, pp. 237–254, 1995.
61. R. Felicetti and P. G. Gambarova, "Effects of high temperature on the residual compressive strength of high-strength siliceous concretes," ACI Materials Journal, vol. 95, no. 4, pp. 395–406, 1998.
62. K. K. Sideris, "Mechanical characteristics of self-consolidating concretes exposed to elevated temperatures," Journal of Materials in Civil Engineering, vol. 19, no. 8, pp. 648–654, 2007.
63. H. Fares, S. Remond, A. Noumowe, and A. Cousture, "High temperature behaviour of self-consolidating concrete. Microstructure and physicochemical properties," Cement and Concrete Research, vol. 40, no. 3, pp. 488–496, 2010.
64. A. Behnood and M. Ghandehari, "Comparison of compressive and splitting tensile strength of high-strength concrete with and without polypropylene fibers heated to high temperatures," Fire Safety Journal, vol. 44, no. 8, pp. 1015–1022, 2009.
65. G. G. Carette, K. E. Painter, and V. M. Malhotra, "Sustained high temperature effect on concretes made with normal portland cement, normal portland cement and slag, or normal portland cement and fly ash," Concrete International, vol. 4, no. 7, pp. 41–51, 1982.
66. R. Felicetti, P. G. Gambarova, G. P. Rosati, F. Corsi, and G. Giannuzzi, "Residual mechanical properties of high-strength concretes subjected to high-temperature cycles," in Proceedings of the International Symposium of Utilization of High-Strength/High-Performance Concrete, pp. 579–588, Paris, France, 1996.
67. J. A. Purkiss, "Steel fibre reinforced concrete at elevated temperatures," International Journal of Cement Composites and Lightweight Concrete, vol. 6, no. 3, pp. 179–184, 1984.

68. P. Rossi, "Steel fiber reinforced concretes (SFRC): an example of French research," ACI Materials Journal, vol. 91, no. 3, pp. 273–279, 1994.
69. V. R. Kodur, "Fibre-reinforced concrete for enhancing structural fire resistance of columns," Fibre-Structural Applications of Fibre-Reinforced Concrete, ACI SP-182, pp. 215–234, 1999.
70. C. R. Cruz, "Elastic properties of concrete at high temperatures," Journat of the PCA Research and Development Laboratories, vol. 8, pp. 37–45, 1966.
71. I. D. Bennetts, Tech. Rep. MRL/PS23/81/001, BHP Melbourne Research Laboratories, Clayton, Australia, 1981.
72. C. Castillo and A. J. Durrani, "Effect of transient high temperture on high-strength concrete," ACI Materials Journal, vol. 87, no. 1, pp. 47–53, 1990.
73. F. P. Cheng, V. K. R. Kodur, and T. C. Wang, "Stress-strain curves for high strength concrete at elevated temperatures," Tech. Rep. NRCC-46973, National Research Council of Canada, 2004.
74. Y. F. Fu, Y. L. Wong, C. S. Poon, and C. A. Tang, "Stress-strain behaviour of high-strength concrete at elevated temperatures," Magazine of Concrete Research, vol. 57, no. 9, pp. 535–544, 2005.
75. U. Schneider, "Concrete at high temperatures—a general review," Fire Safety Journal, vol. 13, no. 1, pp. 55–68, 1988.
76. N. Raut, Response of high strength concrete columns under fire-induced biaxial bending [Ph.D. thesis], Michigan State University, East Lansing, Mich, USA, 2011.
77. Y.-F. Fu, Y.-L. Wong, C.-S. Poon, C.-A. Tang, and P. Lin, "Experimental study of micro/macro crack development and stress-strain relations of cement-based composite materials at elevated temperatures,"Cement and Concrete Research, vol. 34, no. 5, pp. 789–797, 2004.
78. G. A. Khoury, B. N. Grainger, and P. J. E. Sullivan, "Strain of concrete during fire heating to 600°C,"Magazine of Concrete Research, vol. 37, no. 133, pp. 195–215, 1985.
79. J. C. MareÂchal, ACI SP 34, American Concrete Institute, Detroit, Mich, USA, 1972.

80. H. Gross, "High-temperature creep of concrete," Nuclear Engineering and Design, vol. 32, no. 1, pp. 129–147, 1975.
81. U. Schneider, U. Diedrichs, W. Rosenberger, and R. Weiss, Sonderforschungsbereich 148, Arbeitsbericht 1978–1980, Teil II, B 3, Technical University of Braunschweig, Germany, 1980.
82. Y. Anderberg and S. Thelandersson, "Stress and deformation characteristics of concrete at high temperatures, 2-Experimental investigation and material behaviour model," Bulletin 54, Lund Institute of Technology, Lund, Sweden, 1976.
83. V. K. R. Kodur, "Spalling in high strength concrete exposed to fire—concerns, causes, critical parameters and cures," in Proceedings of the ASCE Structures Congress: Advanced Technology in Structural Engineering, pp. 1–9, May 2000.
84. U. Diederichs, U. Jumppanen, and U. Schneider, "High temperature properties and spalling behaviour of high strength concrete," in Proceedings of the 4th Weimar Workshop on High Performance Concrete, HAB, Weimar, Germany, 1995.
85. K. D. Hertz, "Limits of spalling of fire-exposed concrete," Fire Safety Journal, vol. 38, no. 2, pp. 103–116, 2003.
86. Y. Anderberg, "Spalling phenomenon of HPC and OC," in Proceedings of the International Workshop on Fire Performance of High Strength Concrete, NIST SP 919, NIST, Gaithersburg, Md, USA, 1997.
87. Z. P. Bažant, "Analysis of pore pressure, thermal stress and fracture in rapidly heated concrete," inProceedings of the International Workshop on Fire Performance of High Strength Concrete, NIST SP 919, pp. 155–164, Gaithersburg, Md, USA, 1997.
88. A. N. Noumowe, R. Siddique, and G. Debicki, "Permeability of high-performance concrete subjected to elevated temperature (600°C)," Construction and Building Materials, vol. 23, no. 5, pp. 1855–1861, 2009.
89. V. Boel, K. Audenaert, and G. De Schutter, "Gas permeability and capillary porosity of self-compacting concrete," Materials and Structures/Materiaux et Constructions, vol. 41, no. 7, pp. 1283–1290, 2008.
90. U. Danielsen, "Marine concrete structures exposed to hydrocarbon fires," Tech. Rep., SINTEF-The Norwegian Fire Research Institute, Trondheim, Norway, 1997.

91. V. R. Kodur and M. A. Sultan, "Structural behaviour of high strength concrete columns exposed to fire," in Proceedings of the International Symposium on High Performance and Reactive Powder Concrete, pp. 217–232, 1998.
92. V. Kodur and R. McGrath, "Fire endurance of high strength concrete columns," Fire Technology, vol. 39, no. 1, pp. 73–87, 2003.
93. V. K. R. Kodur, F.-P. Cheng, T.-C. Wang, and M. A. Sultan, "Effect of strength and fiber reinforcement on fire resistance of high-strength concrete columns," Journal of Structural Engineering, vol. 129, no. 2, pp. 253–259, 2003.
94. L. T. Phan, "Spalling and mechanical properties of high strength concrete at high temperature," inProceedings of the 5th International Conference on Concrete under Severe Conditions: Environment & Loading (CONSEC '07), CONSEC Committee, Tours, France, 2007.
95. A. Noumowé, P. Clastres, G. Debicki, and J. Costaz, "Thermal stresses and water vapor pressure of high performance concrete at high temperature," in Proceedings of the 4th International Symposium on Utilization of High-Strength/High-Performance Concrete, Paris, France, 1996.
96. V. K. R. Kodur, "Fiber reinforcement for minimizing spalling in High Strength Concrete structural members exposed to fire," ACI, Special Publication, Innovations in Fibre-ReinForced Concrete For Value, 216-14, pp. 221–236, 2003.
97. V. K. R. Kodur, "Design solutions for enhancing the fire resistance of high strength concrete columns,"Indian Concrete Journal, vol. 81, no. 10, pp. 9–20, 2007.
98. V. Kodur, Fire Resistance Design Guidelines for High Strength Concrete Columns, National Research Council, Ontario, Canada, 2003.
99. A. Bilodeau, V. M. Malhotra, and G. C. Hoff, "Hydrocarbon fire resistance of high strength normal weight and light weight concrete incorporating polypropylene fibres," in Proceedings of the International Symposium on High Performance and Reactive Powder Concrete, Sherbrooke, Canada, 1998.

100. A. Bilodeau, V. K. R. Kodur, and G. C. Hoff, "Optimization of the type and amount of polypropylene fibres for preventing the spalling of lightweight concrete subjected to hydrocarbon fire," Cement and Concrete Composites, vol. 26, no. 2, pp. 163–174, 2004.
101. D. P. Bentz, "Fibers, percolation, and spelling of high-performance concrete," ACI Structural Journal, vol. 97, no. 3, pp. 351–359, 2000.
102. V. K. R. Kodur and R. McGrath, "Effect of silica fume and lateral confinement on fire endurance of high strength concrete columns," Canadian Journal of Civil Engineering, vol. 33, no. 1, pp. 93–102, 2006.

Chapter 2

Aerogels as Promising Thermal Insulating Materials: An Overview

Prakash C, Thapliyal, and Kirti Singh

Organic Building Materials Group, CSIR-Central Building Research Institute, Roorkee 247667, India

ABSTRACT

Aerogels are solids with high porosity (<100 nm) and hence possess extremely low density (~0.003 g/cm^3) and very low conductivity (~10 mW/mK). In recent years, aerogels have attracted more and more attention due to their surprising properties and their existing and potential applications in wide range of technological areas. An overview of aerogels and their applications as the building envelope components and respective improvements from an energy efficiency perspective including performance is given here. This overview covers thermal insulation properties of aerogels and studies regarding structural features which will be helpful in buildings envelope. The improvements of thermal insulation systems have future prospects of large savings in primary energy consumption. It can be concluded that aerogels have great potential in a wide range of applications as energy efficient insulation, windows, acoustics, and so forth.

INTRODUCTION

Short supply, limited availability, and increasing energy costs all around the world emphasize the need for immediate energy conservation in both oil rich and oil producing countries. An effective way towards saving energy is to improve the thermal insulation of buildings especially in hot climates where the energy demand for cooling by air conditioning is comparatively higher. In addition to the need for energy saving, high insulating materials are further justified by improved comfort levels and increased building life. Thermal characteristics depend largely on the thermal conductivity of the cell walls and the cell matrix, as well as radiation and convection, with the cell matrix being the most significant factor in determining the overall heat transfer characteristics. Thermal properties of some commonly available insulating materials are given in Table 1.

Table 1: Commonly available thermal insulation materials

S. number	Material	R-value (per inch)	Green	Flammable	Remark
1	Mineral wool	R—3.1	Yes	No	Does not melt or support combustion
2	Fibreglass	R—3.1	Yes	No	Does not absorb water
3	Polystyrene (EPS)	R—4	No	Yes	Is difficult to use around imperfections; can become costly

| 4 | Polyurethane foam | R—6.3 | No | Yes | Makes a great sound insulator |
| 5 | Cellulose | R—3.7 | Yes | Yes | Contains the highest amount of recycled content |

Thermal conductivity varies with time due to changes in the composition of the cell matrix. The ambient air and external building surface temperatures in hot climates of Asia and Africa are much higher than in cold climates of a Australia, Europe and America, temperature of 38°C should be taken into account while calculating thermal conductivities at ageing. In addition to the product specific parameter of the change of the thermal conductivity, mean temperature and water absorption are also other important influencing factors.

As per IUPAC, aerogel is defined as a gel comprised of a microporous solid in which dispersed phase is a gas [1]. Aegerter et al. defined aerogels as gels in which the liquid has been replaced with air, with very moderate shrinkage of a solid network [2]. Aerogel is basically a synthetic porous ultralight material derived from a gel, in which the liquid component of the gel has been replaced with a gas; for example, graphene aerogels are so light that they can rest on top of a grass leaf. The combination of high porosity and extremely small pores provides aerogels with their extreme properties: solid with extremely low density and low thermal conductivity [3]. Aerogels are sometimes also known by different names such as frozen smoke, solid smoke, solid air, or blue smoke owing to translucent nature and the way light scatters in the material [4]. Typical structure of an aerogel is shown in Figure 1.

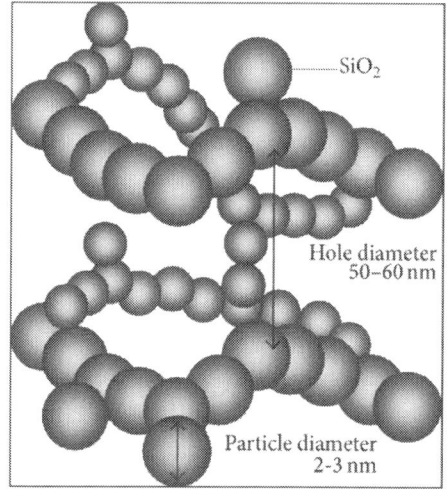

(a)

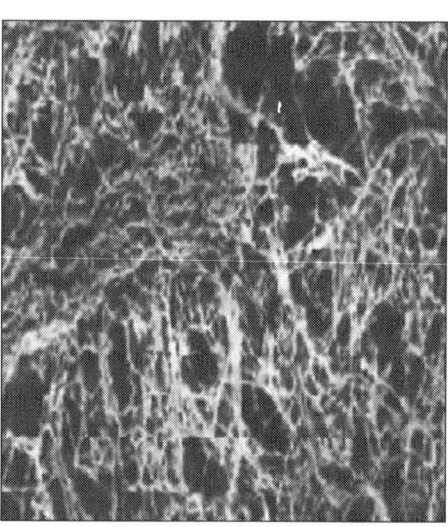

(b)

Figure 1: Nanometer scale particles and pores in an aerogel. (a) Network architecture of an aerogel [77]. (b) Electron micrograph of a silica aerogel [78].

Aerogels together with vacuum insulation panels are one of the new promising high performance thermal and acoustic insulation materials for possible building applications and are currently the main market for aerogels, whereas other applications such as absorbents, shock absorbers, nuclear waste storage, batteries, and catalysts are also possible [5–13]. A list of commercial available aerogels with their trade names is given in Table 2.

Table 2: Commercial aerogel products

S. number	Product	Applications	References
1	Cabot	(i) Pellets, composites	[73, 74]
		(ii) Day lighting applications	
		(iii) Oil and gas pipeline insulation	
		(iv) Cryoinsulation	
2	Aspen aerogels	(i) Construction materials	[74–76]
		(ii) Flexible blanket insulation	
		(iii) Oil and gas pipeline	
		(iv) Aerospace, apparel	
3	Nanopore	(i) Vacuum insulation panels	[74]
		(ii) Shipping containers	
		(iii) Refrigeration	
		(iv) Apparel	

Aerogels are typically characterized by low density solid, low optical index of refraction, low thermal conductivity, low speed of sound through materials, high surface area, and low dielectric constant.

In this paper authors gave an overview of aerogels and their applications as the building envelope components and respective improvements from an energy efficiency perspective. This covers thermal insulation properties of aerogels and studies regarding structural features which will be helpful in buildings envelope. This overview is presented in two parts: firstly, the general discussion of aerogels regarding how they have such a high thermal quality and what are their physical properties which will be useful in making insulating materials and secondly, their remarkable properties due to extraordinary physical and chemical structure of aerogels.

AEROGELS

The passage of thermal energy through an insulating material occurs through three mechanisms: solid conductivity, gaseous conductivity, and radiative (infrared) transmission. The sum of these three components gives the total thermal conductivity of the material. Solid conductivity is an intrinsic property of a specific material. The improvement of thermal resistance of the building envelope can be achieved by decreasing the thermal conductivity.

Fricke et al. observed that both the solid conductivity and the gas conductivity were proportional to the density as shown below:

$$\lambda_{gas} \propto \rho^{-0.6}$$

$$\lambda_{solid} \propto \rho^{1.5}.$$
(1)

Hümmer et al. using these relations derived the following relation for the radiative conductivity, which is a relative equation for the thermal conductivity of opacified silica aerogels:

$$\lambda_{total}(\rho) = \lambda_{solid,0}\left(\frac{\rho}{\rho_0}\right)^{1.5} + \lambda_{gas,0}\left(\frac{\rho}{\rho_0}\right)^{-0.6}$$
$$+ \lambda_{rad,0}\left(\frac{\rho}{\rho_0}\right)^{-1}\left(\frac{T}{T_0}\right)^{3},$$
(2)

where ρ (kg/m³) is the density; λ_{total}, λ_{gas}, λ_{solid}, and λ_{rad} (W/m.K) are the total conductivity, the conductivity for gas conduction, the conductivity for solid conduction, and the radiative conductivity, respectively; T (°K) is the temperature, and the index 0 means that parameters are related to a reference material from an aerogel [14]. Aerogel is made of more than 90% of air, having extremely low weight, transparency, and excellent thermal conductivity. Aerogel is an ideal material for thermal insulation due to all these properties [15, 16]. Also their high visible solar transmittance (T_{vis}) is desirable for application in windows. Further decrease in thermal conductivity of aerogel can be observed if evacuated below 50 hPa; thermal conductivity decreased because of elimination of pore gas. Superinsulations with extremely low thermal conductivities can be implemented with evacuated highly porous powder, fiber, or gel spacers. Due to the Knudsen effect, thermal conductivity can become lower than that for the still air, that is, even less than 25 mW/m.K [17].

For example, silica aerogel is a highly porous material with pore diameters in the range of 10–100 nm. The porosity is more than 90% with a thermal conductivity lower than that of air, which makes these aerogels a highly insulating material. The space not occupied by solids in an aerogel is normally filled with air (or another gas) unless the material is sealed under vacuum. These gases can also transport thermal energy through the aerogel. The pores of silica aerogel are open and allow the passage of gas through the material. The final mode of thermal transport through silica aerogels involves infrared radiation [14]. Soleimani Dorcheh and Abbasi reported the synthesis of nanostructured silicon based transparent aerogels with pore diameter 20–40 nm [18].

Water molecules do not interact strongly with the hydrophobic aerogel pore walls and therefore will not lose much energy in colliding with the wall and the progress of these molecules will not be significantly slowed. Accordingly, the aerogel possesses high breathability, that is, high permeation selectivity between water vapor and agent vapors. Titania aerogels demonstrated an excellent mesoporous structure for application as photoanodes of dye-sensitized solar cells with power conversion efficiency improvement of 16% [19]. Sol-gel derived silica has found tremendous applications as a biocompatible scaffold for the immobilization of cells. A new method for rapid, reproducible, and sensitive detection of rhizobia using aerogels has been reported for the first time [20].

Thermal insulation properties of aerogels are closely related to their acoustic properties too. The acoustic propagation in aerogels depends on the interstitial gas nature and pressure, density, and more generally the texture [21]. Different applications of aerogels are given in Figure 4.

CLASSIFICATION OF AEROGELS

Aerogels can be classified on the basis of the following [22].
- On the basis of appearance
  1. Monolith
  2. Powder
  3. Film/felts
- On the basis of preparation methods
  1. Aerogel
  2. Xerogel
  3. Cryogel
  4. Other aerogel-related materials
- On the basis of different microstructures
  1. Microporous aerogel (<2nm)
  2. Mesoporous aerogel (2-50nm)
  3. Mixed porous aerogel
- On the basis of chemical structure
  1. Oxides
  2. Polymers
  3. Mixed
  4. Hybrid
  5. Composite.

PREPARATION OF AEROGELS

Different type of aerogels can be prepared using alumina, chromium, tin oxide, and carbon, but preparation of silica based aerogel is comparatively easier and reliable. Aerogels are synthesized via a sol-gel process consisting of three main steps [Figure 3].

- *Gel Preparation*: Solid nanoparticles grow crosslink and finally form a three-dimensional solid network with solvent filled pores. To begin with a gel is created in solution and then the liquid is carefully removed to leave the aerogel intact; initially the creation of a colloidal suspension of solid particles known as a "sol" takes place; for example, silica gels are synthesized by hydrolyzing monomeric tetrafunctional and trifunctional silicon alkoxide precursors employing a mineral acid or a base as a catalyst [23, 24]. There are many ways to create silica based sol gels. One is by mixing tetraethoxysilane $Si(OC_2H_5)_4$ with ethanol and water to make it polymerize and thus producing a water based silica gel as shown in (3). A solvent, such as methanol, is used to extract and replace the water [25]:

$$Si(OCH_2C_3)_4 \text{ (liq.)} + 2H_2 \text{ (liq.)}$$
$$\longrightarrow SiO_2 \text{ (solid)} + CH_2CH_3OH \text{ (liq.)} \qquad (3)$$

- *Aging of the Gel*: It provides strength to the structure of gel. The gel prepared earlier is aged in its mother solution [26]. This aging process strengthens the gel, so that minimum shrinkage occurs during the drying step [27]. After gelification, the gel is left undisturbed in the solvent to complete the reaction. After completion of reaction the aerogel product is formed. Inorganic aerogels can be prepared via sol-gel processing, a technique which requires alkoxides or metal salts in alcoholic or aqueous solutions, and subjecting to supercritical drying.
- *Drying:* The solvent has to be removed whilst preserving the solid aerogel network. This can be done either by supercritical drying or at ambient conditions. Aerogel materials are typically prepared by removing the solvent contained in a gel matrix by extraction in a supercritical fluid medium. This can be accomplished by bringing the gel solvent system above its critical temperature and pressure and subsequently relieving pressure above the critical temperature until only vapor remains.

Alternatively, the gel solvent system can be extracted from the wet gel with an appropriate solvent. Liquid carbon dioxide is the most popular extraction solvent because it is inexpensive and has a relatively low critical temperature and critical pressure [28–31].

Crack-free silica aerogels can also be obtained via solvent exchange and resulting surface modification of wet gels using either isopropyl alcohol, trimethylchlorosilane, or n-hexane solution [32]. The physics and chemistry involved in the synthesis of aerogels are detailed in the literature on aerogels [18, 23, 33–37].

For dense silica, solid conductivity is relatively high (a single pane window transmits a large amount of thermal energy). However, silica aerogels possess a very small (~1–10%) fraction of solid silica. Additionally, the solids that are present consist of very small particles linked in a three-dimensional network with many "dead-ends." Therefore, thermal transport through the solid portion of silica aerogel occurs through a very tortuous path and is not particularly effective [38]. Use of methyltrimethoxysilane coprecursor makes the aerogel hydrophobic and makes it able to hold water droplets on the surface [39]. Porosity of silica aerogels was determined by helium pycnometry using the following formula and was found to be 1900 kg/m³ [40]:

$$\text{Porosity}(\%) = \left(1 - \frac{\rho_b}{\rho_s}\right), \tag{4}$$

where P_b is the bulk density and P_s is the skeleton density.

The hydrophobic aerogels have also been obtained via coprecursor method pioneered by Schmidt and Schwertfeger [13]. Hydrolysis and condensation rates of all the coprecursors were observed to be slower than that of TEOS because the former contains one or more nonreactive alkyl/aryl groups, which are nonhydrolysable, and a three-dimensional solid network is achieved as per the following chemical reaction:

$$n\,Si(OC_2H_5)_4 + 2H_2O \xrightarrow[\text{Oxalic Acid}]{C_2H_5OH} SiO_2 + 4n\,C_2H_5OH \tag{5}$$

When sufficient amount of TEOS has hydrolyzed, the silyl groups of the coprecursor get attached to the silica clusters as per the following chemical reactions:

$$\mathrm{-Si}\begin{smallmatrix}\diagup OH\\ -OH\\ \diagdown OH\end{smallmatrix} + (OCH_3)_3Si-CH_3 \longrightarrow -Si\begin{smallmatrix}\diagup O\\ -O\\ \diagdown O\end{smallmatrix}Si-CH_3 + 3CH_3OH$$

$$\equiv Si-OH + Cl-\underset{\underset{CH_3}{|}}{\overset{\overset{H}{|}}{Si}}-CH_3 \longrightarrow \equiv Si-O-\underset{\underset{CH_3}{|}}{\overset{\overset{H}{|}}{Si}}-CH_3 + HCl$$

$$-O-Si\begin{smallmatrix}\diagup OH\\ -OH\\ \diagdown OH\end{smallmatrix} + (C_2H_5O)_3-Si-\bigcirc \longrightarrow -O-Si\begin{smallmatrix}\diagup O\\ -O\\ \diagdown O\end{smallmatrix}Si-\bigcirc + 3C_2H_5OH \tag{6}$$

As the silica clusters get attached to nonhydrolysable organic groups (silyl) on their surfaces, the aerogels become hydrophobic. Hydrophobicity of aerogels will increase with the numbers of alkyl/aryl groups attached to the surface [41].

In case of organic aerogels derived from the sol-gel polymerization of resorcinol with formaldehyde, thermal conductivity components are clearly correlated with the aerogel structure; that is, the solid conductivity can be determined by the porosity and connectivity between the particles while the gaseous conductivity can be influenced by pore size and mass specific infrared absorption of the building units influences radiative transport [42].

Polymer aerogels were prepared from mixtures containing a fixed stoichiometric amount of formaldehyde and varying proportions of resorcinol (RF) and 2,4-dihydroxybenzoic acid (DHBAF) with the objective of combining the advantages of high mesopore volume and solids content of RF aerogels with the ion exchange capacity of DHBAF aerogels, and results show that the aerogel properties vary systematically as the synthesis conditions are changed. It was found that the addition of R to the synthesis mixture resulted in increased values of surface area, mesopore volume, and mean diameter, while simultaneously maintaining the ion exchange capacity of the wet gel [43, 44].

In the TG-DTA of some of the silica aerogel samples there is rapid increase in the weight loss of hydrophilic silica aerogels at 50–100°C due to evaporation of trapped H_2O and alcoholic groups from hydrophilic silica aerogels which were produced by the condensation reactions of Si-OH and $Si(OC_2H_5)$ groups, whereas the percentage of weight loss is negligible up to the temperature of thermal stability in case of hydrophobic aerogels [42].

The effect of heat treatment on the hydrophobicity and specific surface area has been also investigated by several researchers. The results of these studies indicate that hydrophobicity of silica aerogel decreased with increasing the heating temperature to 350°C. On further increasing the heating temperature to 500°C, silica aerogel becomes completely hydrophilic (Figure 2). Some results for MTES coprecursor based aerogels show that the hydrophobicity of the silica around 573 K corresponding to oxidation of aerogel could be maintained up to 350°C [45, 46].

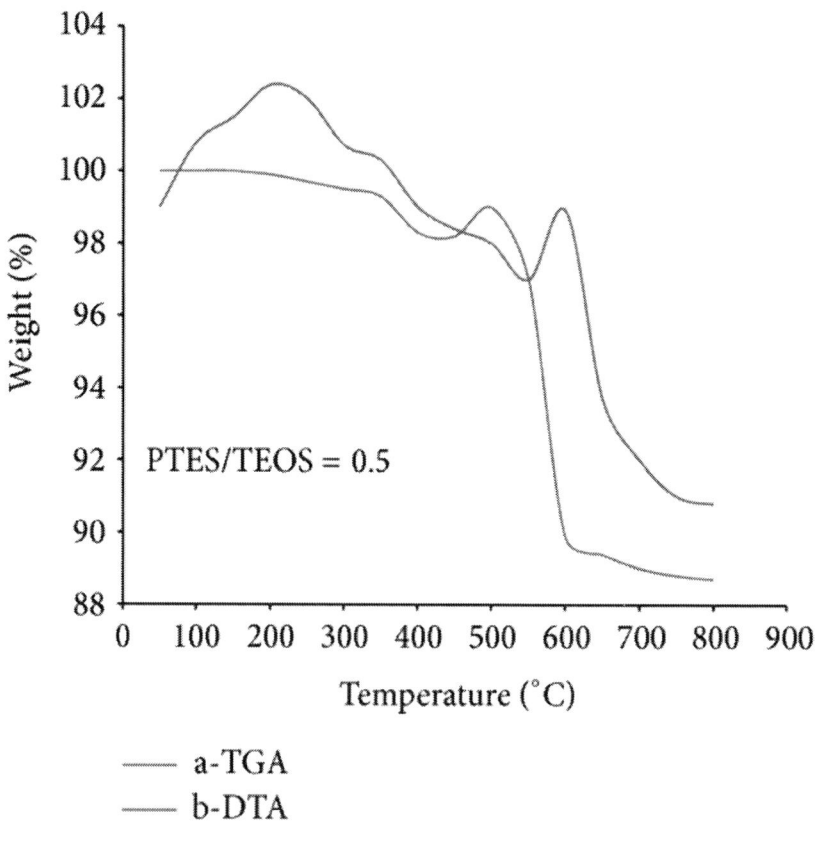

(a)

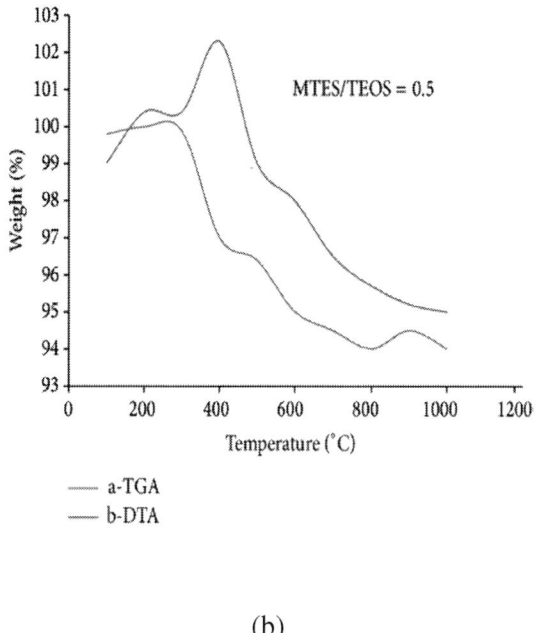

(b)

Figure 2: TGA and DTA analyses of the hydrophobic silica aerogels. (a) PTES/TEOS = 0.5. (b) MTES/TEOS = 0.5 [79].

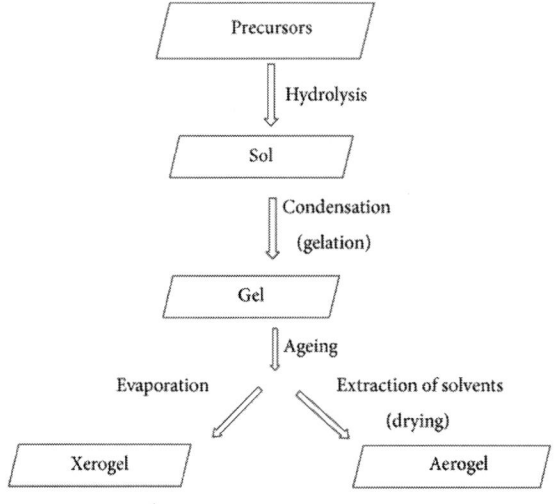

Figure 3: Distinct step of aerogel production [80].

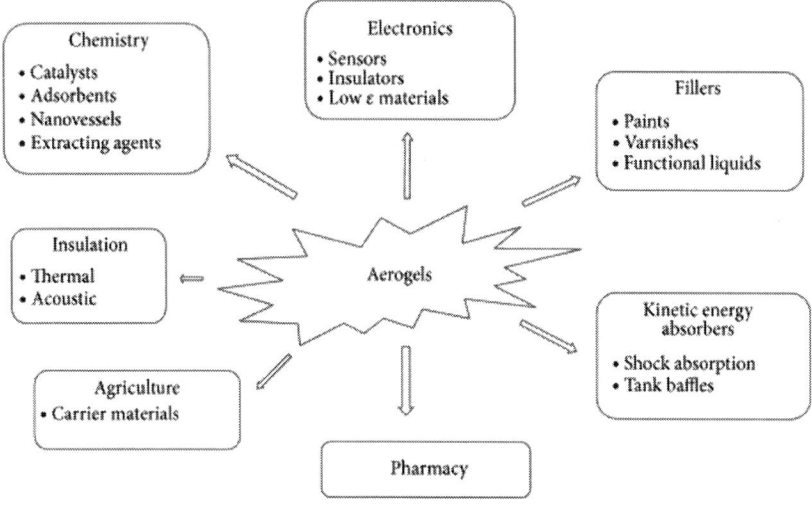

Figure 4: Different applications of aerogels.

STRUCTURAL FEATURES

Aerogels have an unusual combination of high porosity and small pore size, making porosity characterization by conventional techniques, such as mercury intrusion, thermoporometry, and nitrogen adsorption/desorption, very difficult. All these techniques are based on the application of capillary pressures on the aerogel network, which may cause large volumetric compressions, leading to incorrect values for pore size and volume [27]. Aerogels are characterized with a very low permeability which can be explained in terms of pore size suitable for transport of water vapours/gases but not for water molecules [46]. Some aerogels such as carbon aerogels can be obtained in the form of monoliths, beads, powders, or thin films and make them promising materials for application in adsorption and catalysis [47, 48]. Organic polymer aerogels are important nanoporous materials and their nanopore structures can be modified by the chemical reactions. These properties enable carbon nanotube aerogels potential improvement over current carbon aerogels for applications such as sensors, actuators, electrodes, and thermoelectric devices [49]. The porosity provides both molecular accessibility and rapid mass transport via diffusion and for these reasons aerogels have been part of the heterogeneous catalytic materials field for over 50 years. The high porosity and mesoscopic pore diameters of aerogel structures enable the electrolyte to penetrate the entire aerogel particle [50].

An aerogel possesses the following characteristics [22].

- Property characteristics
    1. Ultralow thermal conductivity
    2. Ultralow refractive index
    3. Ultralow dielectric constant
    4. High surface area
    5. High refractive index
    6. Ultralow relative density
    7. Ultrahigh porosity
- Structure characteristics
    1. Gel-like structure on nanoscale coherent skeletons and pores
    2. Hierarchical and fractal microstructure
    3. Macroscopic monolith
    4. Randomly crosslinking network
    5. Noncrystalline matter.

ADVANTAGES OF AEROGELS

Aerogels are regarded as one of the most promising high performance thermal insulation materials for building applications today. With a low thermal conductivity (~13 mW/mK), they show remarkable characteristics compared to traditional thermal insulation materials. Also higher transmittances in the solar spectrum are of great interest for the construction sector. Another advantage of aerogels is their visible transparency for insulation applications which will allow their use in windows and skylights which give architects and engineers the opportunity of reinventing architectural solutions [51]. For example, the low thermal conductivity, a high solar energy, and daylight transmittance in monolithic silica aerogel make it a very interesting material for use in highly energy efficient windows [52]. For cryogenic systems, multilayered insulation (MLI) is the insulation of choice.

However, MLI requires a high vacuum for optimal effectiveness. Powder insulations such as glass microspheres and aerogel beads have shown promise at soft vacuums and have a structural advantage in that they are far simpler to install and maintain [53, 54]. Due to porous structure and low density, aerogels can trap space projectiles travelling with hypervelocity speed (order of km s^{-1}). NASA used aerogel to trap space dust particles and for thermal insulation of space suits [55–57]. One of the promising applications that promote the development of high quality transparent silica aerogel was the use of this low density material in physics as Cherenkov detector [58].

Our indoor environments are polluted by releasing many pollutants like chloride from tap-water, VOCs from organic solvents, formalin from furniture and paints, SO_x and NO_x from incomplete combustion of gasses and many hydrocarbons, and so on. Airborne contaminants are responsible for increasing some respiratory problems and allergies like asthma. The conversion of airborne contaminants into nontoxic compounds is an effective pathway for their removal and to protect our environment. Aerogels can also be used in air purification by removal of airborne contaminants and protect our environment by pollutants [59]. Aerogels are potentially more environmentally friendly than noble metal catalysts due to the negative environmental impact associated with mining and processing the metals [60].

Modification of aerogels is essential to achieve specific functionality and this tailoring can start during the sol-gel process either after gelation or after obtaining the aerogel. This can be done via (a) surface functionalization of aerogels for regulating the adsorption capacity and (b) applying a polymeric coating on aerogel surface. Hybrid aerogels can encompass the intrinsic properties of aerogels (high porosity and surface area) with the mechanical properties of inorganic components and the functionality and biodegradability of biopolymers [61–63].

LIMITATIONS OF AEROGELS

Widespread uses of aerogel materials are restricted at present mainly due to their

- High production costs,
- Poor mechanical properties,
- Health issues.

Supercritical drying is the most expensive and risky aspect of aerogel making process [64]. A highly desirable goal in aerogel preparation is the elimination of the supercritical drying process. For example, Guo and Guadalupe have succeeded in synthesizing a silica based aerogel from a metastable lamellar composite through cooperative interaction between silica and surfactant species [65]. The surfactant molecules used to generate pores can be removed from the silica network through conventional solvent extraction. The porous structure is stable during this procedure, in which no supercritical extraction is used [66]. Silica aerogels are very fragile but strength of silica aerogel monoliths has been improved by a factor of >100 through crosslinking the nanoparticle building blocks of preformed silica hydrogels with poly(hexamethylene diisocyanate). These composite monoliths are much less hygroscopic than native silica, and they do not collapse when in contact with liquids [67].

Aerogels are a mechanical irritant to the eyes, skin, respiratory tract, and digestive system. Small aerogel particles can potentially cause silicosis, and so forth, when inhaled and can induce dryness of the skin, eyes, and mucous membranes. Therefore, protective gear including respiratory protection, gloves, and eye goggles must be worn while handling aerogels [68].

CONCLUSIONS

Paints and coatings can be used for thermal insulation of buildings and work has been done also in CSIR-CBRI [69]. But aerogels are fast becoming alternate material of choice for thermal insulation due to their ultralow thermal conductivity. In the preparation of aerogels supercritical drying is the most effective process. In the conventional preparation of aerogels, expensive raw materials and supercritical drying are used which prohibit commercialization. It is clear that for the large scale commercial aerogel production cost and risk have to be reduced. Aerogels can only be used as building material if we can utilize their highly thermal insulation properties with light weight and low cost.

There is little that can be done to reduce thermal transport through the solid structure of aerogels. Lower density aerogels can be prepared (as low as 0.003 g/cm^3), which reduces the amount of solid present, but this leads to aerogels that are mechanically weaker. Additionally, as the amount of solids decreases the mean pore diameter increases (with an increase in the gaseous component of the conductivity). These are, therefore, generally not suitable for insulation applications. Carbon is an effective absorber of infrared radiation and, in some cases, actually increases the mechanical strength of the aerogel. At ambient pressure the addition of carbon lowers the thermal conductivity from 0.017 to 0.0135 W/mK [Figure 5]. The minimum value for the carbon composite of ~0.0042 W/mK corresponds to ~R30/inch. Hence, conclusion can be drawn that aerogels have great future potential in a wide range of applications as energy efficient insulation, windows, acoustics, and so forth [41, 70, 71].

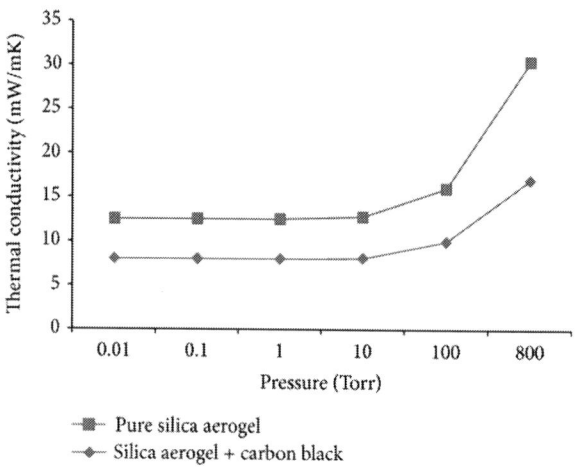

Figure 5: Thermal conductivity versus pressure curves for (i) pure silica aerogel and (ii) single-step silica aerogel with 9% (wt/wt) carbon black [70].

Chiral mesoporous SiO_2 (CMS) as shown in Figure 6 can be synthesized with amino acid block copolymers and their acoustically induced optical Kerr effects (AIOKE) were found very high compared to nonchiral SiO_2 and therefore CMS can be used in acoustically operated quantum electronic devices [72]. Recently cellulose nanofibril (CNF) aerogels with superior wet resilence and water activated shape recovery were fabricated without chemical crosslinking by ice crystal templated self assembly of TEMPO oxidised CNFs via cyclic freezing thawing method. Main challenge lies in the strengthening of aerogels either by crosslinking with cellulosic polymers or the incorporation of cellulose-based nanofibres. Other challenge is lowering the production cost of composite/hybrid aerogel materials via ambient drying and continuous production technology.

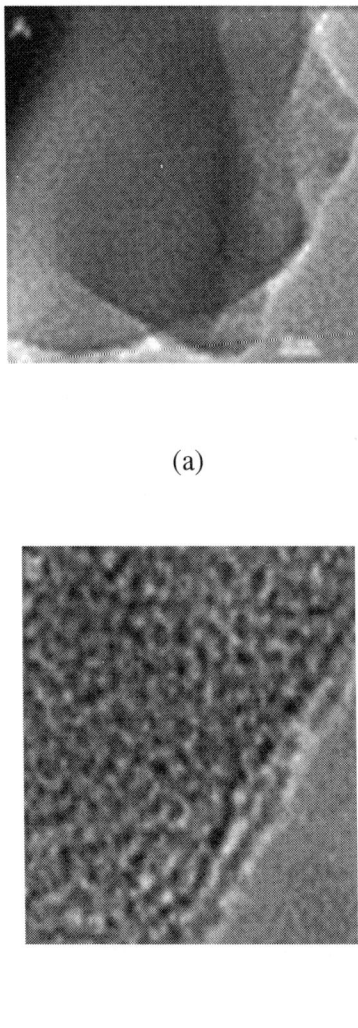

Figure 6: HRTEM images of chiral Ex-SiO$_2$-CBC, (a) at low and (b) at high resolution [72].

ACKNOWLEDGMENT

The authors are grateful to the Director, CSIR-CBRI, for his continuous guidance and encouragement.

REFERENCES

1. A. D. McNaught and A. Wilkinson, Compendium of Chemical Terminology, IUPAC Goldbook, PAC, 2007, 791801, Blackwell Science, Oxford, Cambridge, UK, 2nd edition. ·
2. M. A. Aegerter, N. Leventis, and M. M. Koebel, Aerogels Handbook, Springer, New York, NY, USA, 2011.
3. "Guinness Records Names JPL›s Aerogel World›s Lightest Solid," NASA Jet Propulsion Laboratory, 2002.
4. K. Higa, "Aerogel—the insulative frozen smoke," Illumin, vol. 14, no. 3, p. 1, 2014.
5. S. S. Kistler, "Coherent expanded aerogels and jellies," Nature, vol. 127, no. 3211, p. 741, 1931.
6. D. M. Smith, R. Deshpande, and C. J. Brinker, "Preparation of low-density aerogels at ambient pressure for thermal insulation," Ceramic Transactions, vol. 31, pp. 71–80, 1993.
7. S. T. Mayer, R. W. Pekala, and J. L. Kaschmitter, "Aerocapacitor. An electrochemical double-layer energy-storage device," Journal of the Electrochemical Society, vol. 140, no. 2, pp. 446–451, 1993.
8. N. Leventis, N. Chandrasekaran, C. Sotiriou-Leventis, and A. Mumtaz, "Smelting in the age of nano: iron aerogels," Journal of Materials Chemistry, vol. 19, no. 1, pp. 63–65, 2009. · ·
9. B. C. Tappan, M. H. Huynh, M. A. Hiskey et al., "Ultralow-density nanostructured metal foams: combustion synthesis, morphology, and composition," Journal of the American Chemical Society, vol. 128, no. 20, pp. 6589–6594, 2006. · ·
10. N. Hüsing and U. Schubert, "Aerogels—airy materials: chemistry, structure, and properties," Angewandte Chemie International Edition, vol. 37, no. 1-2, pp. 22–45, 1998.

11. K. Richter, P. M. Norris, and C. L. Chang, "Aerogels: applications, structure and heat transfer phenomena," in Review on Heat Transfer, V. Prasad, Y. Jaluria, and G. Chen, Eds., vol. 6, chapter 2, pp. 61–114, 1995.
12. L. W. Hrubesh, "Aerogel applications," Journal of Non-Crystalline Solids, vol. 225, no. 1–3, pp. 335–342, 1998.
13. M. Schmidt and F. Schwertfeger, "Applications for silica aerogel products," Journal of Non-Crystalline Solids, vol. 225, no. 1–3, pp. 364–368, 1998.
14. J. Fricke, "Thermal transport in porous superinsulations," in Aerogels, J. Fricke, Ed., vol. 6 of Springer Proceedings in Physics, pp. 94–103, 1986.
15. Z. Novak and Ž. Knez, "Diffusion of methanol-liquid CO_2 and methanol-supercritical CO_2 in silica aerogels," Journal of Non-Crystalline Solids, vol. 221, no. 2-3, pp. 163–169, 1997.
16. H. Tamon, T. Kitamura, and M. Okazaki, "Preparation of silica aerogel from TEOS," Journal of Colloid and Interface Science, vol. 197, no. 2, pp. 353–359, 1998. · ·
17. D. M. Smith, A. Maskara, and U. Boes, "Aerogel-based thermal insulation," Journal of Non-Crystalline Solids, vol. 225, no. 1–3, pp. 254–259, 1998.
18. A. Soleimani Dorcheh and M. H. Abbasi, "Silica aerogel; synthesis, properties and characterization,"Journal of Materials Processing Technology, vol. 199, no. 1, pp. 10–26, 2008. · ·
19. Y. C. Chiang, W. Y. Cheng, and S. Y. Lu, "Titania aerogels as a superior mesoporous structure for photoanodes of dye-sensitized solar cells," International Journal of Electrochemical Science, vol. 7, no. 8, pp. 6910–6919, 2012.
20. P. Arora, S. Sharma, S. K. Ghoshal, N. Dilbaghi, and A. Chaudhury, "A functional approach toward xerogel immobilization for encapsulation biocompatibility of Rhizobium toward biosensor," Frontiers in Biology, vol. 8, no. 6, pp. 626–631, 2013. ·
21. L. Forest, V. Gibiat, and T. Woignier, "Biot›s theory of acoustic propagation in porous media applied to aerogels and alcogels," Journal of Non-Crystalline Solids, vol. 225, no. 1–3, pp. 287–292, 1998.

22. A. Du, B. Zhou, Z. Zhang, and J. Shen, "A special material or a new state of matter: a review and reconsideration of the aerogel," Materials, vol. 6, no. 3, pp. 941–968, 2013. ·
23. C. J. Brinker, R. Sehgal, S. L. Hietala et al., "Sol-gel strategies for controlled porosity inorganic materials," Journal of Membrane Science, vol. 94, pp. 85–102, 1994. ·
24. J. L. Gurav, D. Y. Nadargi, and A. V. Rao, "Effect of mixed Catalysts system on TEOS-based silica aerogels dried at ambient pressure," Applied Surface Science, vol. 255, no. 5, pp. 3019–3027, 2008. · ·
25. A. Acharya, D. Joshi, and V. A. Gokhale, "AEROGEL—a promising building material for sustainable buildings," Chemical and Process Engineering Research, vol. 9, pp. 1–6, 2013.
26. I. Smirnova, S. Suttiruengwong, and W. Arlt, "Feasibility study of hydrophilic and hydrophobic silica aerogels as drug delivery systems," Journal of Non-Crystalline Solids, vol. 350, pp. 54–60, 2004. · ·
27. R. P. Patel, N. S. Purohit, and A. M. Suthar, "An overview of silica aerogels," International Journal of ChemTech Research, vol. 1, no. 4, pp. 1052–1057, 2009.
28. A. C. Pierre and G. M. Pajonk, "Chemistry of aerogels and their applications," Chemical Reviews, vol. 102, no. 11, pp. 4243–4265, 2002. · ·
29. Y. K. Akimov, "Fields of application of aerogels (review)," Instruments and Experimental Techniques, vol. 46, no. 3, pp. 287–299, 2003. · ·
30. G. M. Pajonk, "Some applications of silica aerogels," Colloid and Polymer Science, vol. 281, no. 7, pp. 637–651, 2003. · ·
31. A. Bisson, A. Rigacci, D. Lecomte, E. Rodier, and P. Achard, "Drying of silica gels to obtain aerogels: phenomenology and basic principles," Drying Technology, vol. 21, no. 4, pp. 593–628, 2003. ·
32. C. J. Lee, G. S. Kim, and S. H. Hyun, "Synthesis of silica aerogels from waterglass via new modified ambient drying," Journal of Materials Science, vol. 37, no. 11, pp. 2237–2241, 2002. · ·
33. A. C. Pierre, "The chemistry of precursors solutions," in Introduction to Sol-Gel Processing, vol. 1, chapter 2, pp. 11–89, 1998.

34. A. P. Ambekar and P. Bagade, "A review on: aerogel—world's lightest solid," Popular Plastics & Packaging, vol. 51, pp. 96–102, 2006.
35. A. C. Pierre and A. Rigacci, "SiO_2 aerogels," in Aerogels Handbook, M. A. Aegerter, N. Leventis, and M. M. Koebel, Eds., Advances in Sol-Gel Derived Materials and Technologies, pp. 21–45, 2011.
36. C. E. Carraher Jr., "General topics: silica aerogels—properties and uses," Polymer News, vol. 30, no. 12, pp. 386–388, 2005. ·
37. C. E. Carraher Jr., "Silica aerogels—synthesis and history," Polymer News, vol. 30, pp. 62–64, 2005. ·
38. G. M. Pajonk, "Transparent silica aerogels," Journal of Non-Crystalline Solids, vol. 225, no. 1-3, pp. 307–314, 1998.
39. A. Venkateswara Rao and G. M. Pajonk, "Effect of methyltrimethoxysilane as a co-precursor on the optical properties of silica aerogels," Journal of Non-Crystalline Solids, vol. 285, no. 1–3, pp. 202–209, 2001. · ·
40. A. V. Rao and R. R. Kalesh, "Comparative studies of the physical and hydrophobic properties of TEOS based silica aerogels using different co-precursors," Science and Technology of Advanced Materials, vol. 4, no. 6, pp. 509–515, 2003. · ·
41. X. Lu, R. Caps, J. Fricke, C. T. Alviso, and R. W. Pekala, "Correlation between structure and thermal conductivity of organic aerogels," Journal of Non-Crystalline Solids, vol. 188, no. 3, pp. 226–234, 1995.
42. S.-K. Kang and S.-Y. Choi, "Synthesis of low-density silica gel at ambient pressure: effect of heat treatment," Journal of Materials Science, vol. 35, no. 19, pp. 4971–4976, 2000. · ·
43. R. W. Pekala and F. M. Kong, "Resorcinol-formaldehyde aerogels and their carbonized derivatives,"Polymer Preprints, vol. 30, no. 1, pp. 221–223, 1989.
44. P. J. M. Carrott, L. M. Marques, and M. M. L. R. Carrott, "Core-shell polymer aerogels prepared by co-polymerisation of 2,4-dihydroxybenzoic acid, resorcinol and formaldehyde," Microporous and Mesoporous Materials, vol. 158, pp. 170–174, 2012. · ·

45. F. Shi, L. Wang, J. Liu, and M. Zeng, "Effect of heat treatment on silica aerogels prepared via ambient drying," Journal of Materials Science and Technology, vol. 23, no. 3, pp. 402–406, 2007.
46. B. Hosticka, P. M. Norris, J. S. Brenizer, and C. E. Daitch, "Gas flow through aerogels," Journal of Non-Crystalline Solids, vol. 225, no. 1–3, pp. 293–297, 1998.
47. C. Moreno-Castilla and F. J. Maldonado-Hódar, "Carbon aerogels for catalysis applications: an overview," Carbon, vol. 43, no. 3, pp. 455–465, 2005. · ·
48. M. B. Bryning, D. E. Milkie, M. F. Islam, L. A. Hough, J. M. Kikkawa, and A. G. Yodh, "Carbon nanotube aerogels," Advanced Materials, vol. 19, no. 5, pp. 661–664, 2007. · ·
49. Y. Tao, M. Endo, and K. Kaneko, "A review of synthesis and nanopore structures of organic polymer aerogels and carbon aerogels," Recent Patents on Chemical Engineering, vol. 1, pp. 192–200, 2008.
50. W. Dong, J. S. Sakamoto, and B. Dunn, "Electrochemical properties of vanadium oxide aerogels,"Science and Technology of Advanced Materials, vol. 4, no. 1, pp. 3–11, 2003. · ·
51. R. Baetens, B. P. Jelle, and A. Gustavsen, "Aerogel insulation for building applications: a state-of-the-art review," Energy and Buildings, vol. 43, no. 4, pp. 761–769, 2011. · ·
52. K. I. Jensen, "Passive solar component based on evacuated monolithic silica aerogel," Journal of Non-Crystalline Solids, vol. 145, pp. 237–239, 1992.
53. J. E. Fesmire, S. D. Augustynowicz, and S. Rouanet, "Aerogel beads as cryogenic thermal insulation system," in Proceedings of the Cryogenic Engineering Conference (CEC ‹01), vol. 613 of Advances in Cryogenic Engineering, pp. 1541–1548, July 2001. ·
54. A. L. Nayak and C. L. Tien, "Thermal conductivity of microsphere cryogenic insulation," Advances in Cryogenic Engineering, vol. 22, pp. 251–262, 1977.
55. R. Wawryk and J. Rafalowicz, "The influence of residual gas pressure on the thermal conductivity of microsphere insulations," International Journal of Thermophysics, vol. 9, no. 4, pp. 611–625, 1988. · ·

56. M. J. Burchell, M. J. Cole, M. C. Price, and A. T. Kearsley, "Experimental investigation of impacts by solar cell secondary ejecta on silica aerogel and aluminum foil: implications for the Stardust Interstellar Dust Collector," Meteoritics and Planetary Science, vol. 47, no. 4, pp. 671–683, 2012. · ·
57. J. E. Fesmire, "Aerogel insulation systems for space launch applications," Cryogenics, vol. 46, no. 2-3, pp. 111–117, 2006. · ·
58. M. Tabata, I. Adachi, Y. Ishii, H. Kawai, T. Sumiyoshi, and H. Yokogawa, "Development of transparent silica aerogel over a wide range of densities," Nuclear Instruments and Methods in Physics Research A: Accelerators, Spectrometers, Detectors and Associated Equipment, vol. 623, no. 1, pp. 339–341, 2010. · ·
59. N. J. H. Dunna, M. K. Carrolla, and A. M. Anderson, "Characterization of alumina and nickel-alumina aerogels prepared via rapid supercritical extraction," Polymer Preprints, vol. 52, no. 1, pp. 250–251, 2011.
60. R. Yang, Y.-P. Zhang, and R.-Y. Zhao, "An improved model for analyzing the performance of photocatalytic oxidation reactors in removing volatile organic compounds and its application," Journal of the Air and Waste Management Association, vol. 54, no. 12, pp. 1516–1524, 2004.
61. K. Kanamori, "Organic-inorganic hybrid aerogels with high mechanical properties via organotrialkoxysilane-derived sol-gel process," Journal of the Ceramic Society of Japan, vol. 119, no. 1385, pp. 16–22, 2011.
62. J. L. Plawsky, H. Littman, and J. D. Paccione, "Design, simulation, and performance of a draft tube spout fluid bed coating system for aerogel particles," Powder Technology, vol. 199, no. 2, pp. 131–138, 2010. · ·
63. H. Ramadan, T. Coradin, S. Masse, and H. El-Rassy, "Synthesis and characterization of mesoporous hybrid silica-polyacrylamide aerogels and xerogels," Silicon, vol. 3, no. 2, pp. 63–75, 2011. · ·
64. J. Fricke and A. Emmerling, "Aerogels—recent progress in production techniques and novel applications," Journal of Sol-Gel Science and Technology, vol. 13, no. 1–3, pp. 299–303, 1999.

65. Y. Guo and A. R. Guadalupe, "Functional silica aerogel from metastable lamellar composite," Chemical Communications, no. 4, pp. 315–316, 1999.
66. S. Dai, Y. H. Ju, H. J. Gao, J. S. Lin, S. J. Pennycook, and C. E. Barnes, "Preparation of silica aerogel using ionic liquids as solvents," Chemical Communications, no. 3, pp. 243–244, 2000.
67. N. Leventis, C. Sotiriou-Leventis, G. Zhang, and A.-M. M. Rawashdeh, "Nanoengineering strong silica aerogels," Nano Letters, vol. 2, no. 9, pp. 957–960, 2002.
68. Cryogel 5201, 10201 Safety Data Sheet, Aspen Aerogels. 11/13/07.
69. P. C. Thapliyal and S. R. Karade, "Studies on physico-mechanical behavior of thermal insulating coatings for buildings," in Proceedings of the International Conference on Advanced Materials for Energy Efficient Buildings (AME2B ‹13), no. TS4, p. 52, 2013.
70. T. Rettlebach, J. Sauberlich, S. Korder, and J. Fricke, "Thermal conductivity of silica aerogel powders at temperatures from 10 to 275K," Journal of Non-Crystalline Solids, vol. 186, pp. 278–284, 1995.
71. J. V. Accorsi, "The impact of carbon black morphology and dispersion on the weatherability of polyethylene," KGK-Kautschuk und Gummi Kunststoffe, vol. 54, no. 6, pp. 321–326, 2001.
72. P. Paik, Y. Mastai, I. Kityk, P. Rakus, and A. Gedanken, "Synthesis of amino acid block-copolymer imprinted chiral mesoporous silica and its acoustically-induced optical Kerr effects," Journal of Solid State Chemistry, vol. 192, pp. 127–131, 2012.
73. http://www.cabot-corp.com.
74. M. A. B. Meador, B. N. Nguyen, H. Guo et al., "Aerogels: thinner, lighter, stronger," 2011, www.nasa.gov/topics/technology/features/aerogels.html.
75. Aspen Aerogels, "Insulated building materials," US Patent 7771609, 2010.
76. Aspen Aerogels, "Insulated building materials," US Patent 8277676, 2012.
77. http://www.aerogel.com.

78. 2011, http://www.chem-eng.kyushu-u.ac.jp/e/research.html.
79. A.-Y. Jeong, S.-M. Koo, and D.-P. Kim, "Characterization of hydrophobic SiO_2 powders prepared by surface modification on wet gel," Journal of Sol-Gel Science and Technology, vol. 19, no. 1–3, pp. 483–487, 2000. · ·
80. T. Błaszczyński, A. Ślosarczyk, and M. Morawski, "Synthesis of aerogel by supercritical drying method,"Procedia Engineering, vol. 57, pp. 200–206, 2013. ·

Chapter 3

Atmospheric Corrosion of Painted Galvanized and 55%Al-Zn Steel Sheets: Results of 12 Years of Exposure

C. I. Elsner[1,2] P. R. Seré[1], and A. R. Di Sarli[1]

[1]CIDEPINT, Centro de Investigación y Desarrollo en Tecnología de Pinturas (CICPBA-CCT CONICET-La Plata), Avenida 52 s/n entre 121 y 122, B1900AYB La Plata, Argentina

[2]Facultad de Ingeniería, Universidad Nacional de La Plata, Avenida 1 esq. 47, B1900TAG La Plata, Argentina

ABSTRACT

Zinc or 55%Al-Zn alloy-coated steel sheets, either bare or covered by different painting systems, have been exposed for 12 years to the action of the urban atmosphere at the CIDEPINT station located in La Plata (34° 50′ South, 57° 53′, West), province of Buenos Aires, Argentina. The samples exposed surface was evaluated through periodical visual inspections, standardized adhesion tests, and electrochemical impedance measurements. The ambient variables monitored were average annual rains and temperatures, time of wetness, sulphur and chloride concentration, relative humidity, and speed and direction of the winds. It was found that in this atmosphere, the corrosion resistance of the bare 55% Al-Zn/steel sheets was higher than of the galvanized steel, and the polyurethane painting system was more protective than the alkyd and epoxy ones, which degraded after 6-7 years of exposure.

INTRODUCTION

Exposed to specific aggressive media, metal or alloy stability depends upon the protective properties of the surface film formed, because its chemical composition, conductivity, adherence, solubility, hygroscopicity, and morphological characteristics determine the film capacity to work as a controlling barrier [1]. In such a sense, steel galvanic protection by means of zinc or zinc alloys is a common example, owing not only to the fact that the zinc, being electrochemically more active than the steel, corrodes preferentially, but also to the barrier effect of the corrosion products precipitated on the metallic surface. In particular, the coatings based on zinc are widely used to protect steel structures against atmospheric corrosion [2], because of the protective properties afforded by an insoluble film of basic carbonate. However, if the exposure conditions are such that there is changes of the ambient variables like atmospheric conditions, UV radiation intensity, type and level of pollutants, wet-dry cycles, depletion of air but high humidity, or a medium containing strongly aggressive species like chloride or sulphate ions, the zinc could dissolve forming soluble, less dense, and scarcely protective corrosion products, which sometimes lead to localized corrosion [2–5]. This condition can be reached during the storage and transportation of galvanized steel sheets or when they are exposed to marine and/or industrial environments [6]. Aluminium coatings have overcome these two factors. Nevertheless, as they cannot provide cathodic protection to exposed steel in most environments, early rusting occurs at coating defects and cut edges; besides, these coatings are also subjected to crevice corrosion in marine environmental [7].

For years, many attempts to improve the corrosion resistance of zinc and aluminium coatings through alloying were carried out. Although the protective effect of combinations of these two elements was known, they were not used until the discovery that silicon inhibits the fast alloying reaction with steel [8]. Thus, the alloy commercially known as Galvalume or Zincalum arose, and its composition: 55% Al, 1.6% Si, the rest zinc, was selected from a systematic study, providing an excellent combination of galvanic protection and low corrosion rate.

When a higher degree of protection of these metallic surfaces is of concern, properly chosen painting systems can provide a more effective corrosion-inhibiting barrier and also a better aesthetic appearance [9]. Some exposure conditions are so aggressive that both protective systems (metallic + organic coatings) must be applied to get longer effectiveness. Such a combination, referred to as a duplex system, has demonstrated a synergistic effect when compared to the individual coating systems. This better corrosion protection is attributed to the double action afforded by the zinc or 55%Al-Zn layer (cathodic protection + blockage of its defects by the corrosion products), and also by the pigmented paint system (barrier effect + steel corrosion-inhibition) [10]. Besides, this duplex system requires less reconditioning and repairs of coating systems after transportation and assembly on site.

The mechanism responsible for the protective action of paint coatings is highly complex because it depends upon the simultaneous action of different factors. Irrespective of their intended function (functional, decorative, or protective), the paint must adhere satisfactorily to the underlying substrate [11, 12].

The organic film permeability is important in metallic substrate corrosion, since this property is directly connected to the permeation of environmental corrosion-inducing chemicals through the polymeric matrix, the chemical composition of the latter, and the presence of pores, voids, or other defects in the coating. It is important to note that particularly water and oxygen can permeate the film, at least to some extent, even if none of the intrinsic structural defects are present. For these reasons, the painted metal's resistance to degradation produced by weathering is a very important variable, since it defines the material's durability. Within this concept, the corrosion and resistance to weathering (degradation due to UV radiation, oxygen, humidity, etc.) are separately evaluated. The corrosion resistance depends upon the permeability (barrier effect) of the primer and galvanic layer as well as of the inhibitive capacity of the contained anticorrosive pigments. The weathering degradation (loss of gloss and/or adhesion, chalking, cracking, blistering, etc.) takes place at shorter times and depends mainly on the topcoat paint properties.

The still unsolved paint delamination or blistering problem, due to a bad bond at the substrate/paint interface, depends upon the chemical nature and crosslinking degree of the polymer as well as the metal substrate and its surface treatment [13]. In principle, paint adhesion can be improved by providing the substrate with a pretreatment layer, followed by applying a corrosion inhibiting primer + intermediate and/or topcoat paints. In line with this definition, the primer is considered the critical element in most paint systems because it is mainly responsible for preserving the metallic state of the substrate, and it must also anchor the total paint coating to the steel. Most coatings adhere to the metal via purely physical attractions (e.g., hydrogen bonds) that develop when two surfaces are brought closely together [14, 15]. Paint vehicles with polar groups (–OH, –COOH, etc.) have good wetting properties and show excellent physical adhesion characteristics (epoxies, alkyds, oil paints, etc.). Much stronger chemically bonded adhesion is possible when the primer can actually react with the metal, as is the case of several pretreatments [16–18].

Paint life depends on several factors such as the metallic substrate, the selected paint system, and the paint-substrate interface [19]. Paint selection is generally based on the aggressive medium properties, while the metal treatment before painting has a substantial impact on the useful life of the selected system.

The susceptibility to degradation of painted metals is estimated by accelerated laboratory tests and natural atmospheric exposure for several years [20–31]. Although the extrapolation of accelerated test results do not compare linearly to the actual performance of the coatings in their service life, it can supply useful information related to the rate and form of the corrosion-inhibiting system degradation. In most cases, such information can help to improve the paint formulation and/or the painting scheme design. Consequently, a comparative evaluation of the protective performance of either bare or covered with three different painting systems is reported in this paper. The corrosion resistance of these samples was tested by exposure for 12 years to the action of the urban atmosphere at the CIDEPINT station located in La Plata (34° 50' South, 57° 53', West), province of Buenos Aires, Argentina.

The evolution of the samples exposed surface was evaluated through periodical visual inspections, adhesion tests according to the ASTM D-3359/09 standard, and electrochemical impedance measurements applied to samples immersed for 1 h in 0.5 M Na_2SO_4 solution. The ambient variables monitored were average annual rains and temperatures, time of wetness, sulphur and chloride concentration, relative humidity, and speed and direction of the winds.

EXPERIMENTAL DETAILS

A total of 280 commercial-grade steel sheets (15 × 8 × 0.2 cm) hot-dip coated with zinc or 55%Al-Zn plates were used as the metallic substrate. They were degreased by immersion in 5% Na_2CO_3 solution and then rinsed with distilled water to eliminate any possible surface contamination.

The commercial-grade protective painting systems (Table 1) were applied by brushing to maintain the same conditions for all the samples. After applying the painting system, the painted plates were placed in a dessicator cabinet at controlled temperature (30 ± 2 °C) until completely dry. Next, measurements of dry film thickness (Table 1) were taken with an Elcometer 300 coating thickness gauge, using a bare sanded plate and standards of known thickness as references. Plates were exposed at 45° from the horizontal and they were oriented east allowing a maximum insolation on them. The test site was formed by wooden structures, where the flat plates were set on. Location coordinates meteorological data, atmospheric pollutants, time of wetness and corrosion category of the place of location according to ISO 9223 is given in Figure 1 and Tables 2 and 3, respectively. All environmentally data were available from the Seismology and Meteorological Department of the Astronomic and Geophysics Science Faculty of the National University of La Plata, placed very near CIDEPINT Station.

Table 1: Mean thicknesses (μm)

Metal/paint system	Metallic coating	Primer	Topcoat	Total thickness
S/Z/AS	18 ± 0.9	22.± 0.9	52.± 2.6	92 ± 4.5
S/ZA/AS	20 ± 0.9	22 ± 0.9	52.± 2.6	94 ± 4.5
S/Z/ES	18 ± 0.9	4± 0.2	87.± 2.6	109 ± 5.1
S/ZA/ES	20 ± 0.9	4± 0.2	87.± 2.6	111 ± 5.1
S/Z/PS	18 ± 0.9	4± 0.2	48.± 2.6	70 ± 3.5
S/ZA/PS	20 ± 0.9	4± 0.2	48.± 2.6	72 ± 3.5

Note: S/Z/AS: galvanized steel/alkyd-based paint system; S/ZA/AS: steel/55%Al-Zn alloy/alkyd-based paint system; S/Z/ES: galvanized steel/epoxy-based paint system; S/ZA/ES: steel/55%Al-Zn alloy/epoxy-based paint system; S/Z/PS: galvanized steel/polyurethane-based paint system; S/ZA/PS: steel/55%Al-Zn alloy/polyurethane-based paint system.

Table 2: Meteorological data for the 12 years of exposure in the CIDEPINT Station, La Plata, Argentina

Year	Mean temperature (°C)	Mean relative humidity (%)	Precipitation (mm)	Days of rain
1	16.5	80.8	943.1	90.0
2	16.8	80.7	1042.8	94.0
3	16.3	80.4	927.0	93.0
4	15.8	81.8	1342.1	90.0
5	16.4	84.4	1316.6	98.0
6	16.0	82.7	1611.8	107.0
7	15.8	81.1	924.5	98.0
8	16.1	80.8	881.7	88.0
9	15.9	79.4	926.6	91.0
10	16.0	79.0	1083.4	81.0
11	15.4	77.5	1153.2	85.0
12	16.3	78.3	775.1	70.0

Table 3: Average levels of chemical agents, time of wetness and corrosion category of CIDEPINT Station

Year	Deposition rate of SO_2 (mg·m^{-2}·d^{-1})	Deposition rate of chloride (mg·m^{-2}·d^{-1})	Time of wetness fraction	Corrosion category according to ISO 9223
1	6.22	Negligible	0.61	$P_0S_{0\tau_4}/C_2$
2	7.25	Negligible	0.62	$P_0S_{0\tau_4}/C_2$
3	6.81	Negligible	0.61	$P_0S_{0\tau_4}/C_2$
4	6.53	Negligible	0.65	$P_0S_{0\tau_5}/C_2$
5	7.42	Negligible	0.63	$P_0S_{0\tau_5}/C_2$
6	6.94	Negligible	0.69	$P_0S_{0\tau_5}/C_2$
7	7.60	Negligible	0.59	$P_0S_{0\tau_4}/C_2$
8	8.12	Negligible	0.57	$P_0S_{0\tau_4}/C_2$
9	7.93	Negligible	0.61	$P_0S_{0\tau_4}/C_2$
10	6.82	Negligible	0.62	$P_0S_{0\tau_4}/C_2$
11	7.45	Negligible	0.62	$P_0S_{0\tau_4}/C_2$
12	8.01	Negligible	0.55	$P_0S_{0\tau_4}/C_2$

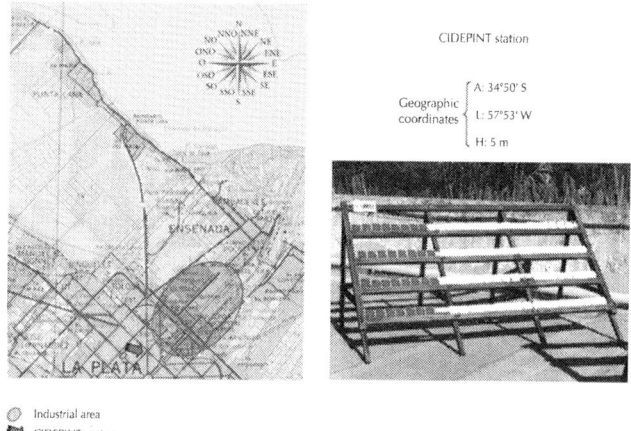

Figure 1: Location of the CIDEPINT station in La Plata, Buenos Aires, Argentina.

To check reproducibility, a total of 280 samples including bare or painted steel/zinc or steel/55%Al-Zn sheets were exposed to natural weathering in La Plata station for 12 years. The painted samples were build-up with the three painting systems mentioned in Table 1, and their edges were masked with a thick wax base coating to avoid edge effect.

Gravimetric determinations for measuring weight-loss of bare steel/metallic coating samples were carried out in triplicate for each material tested.

The visual inspections and samplings took place according to the following program: during the first year each 1st, 3rd, 6th, 9th, 12th month, and then each 2nd, 4th, 8th, and 12th year. At the same times, adhesion tests according to the Test Tape ASTM D-3359/09 Standard on replicates of each type of painted samples were also performed.

Electrochemical Measurements

For the impedance measurements periodically carried out on other replicates of each type of samples, a cylindrical clamp-on acrylic (polymethyl methacrylate) cell was positioned on the painted panel by an O-ring defining a surface area of $15.9\,cm^2$. An aperture in the top of this three electrode electrochemical cell contained a Pt-Rh mesh counter-electrode with negligible impedance, oriented parallel to the working electrode (painted metal surface). A glass-linear Saturated Hg/HgSO$_4$ tipped Reference Electrode was positioned, together with the counter-electrode, close to the exposed painted steel surface panel. For further easy comparison with previous information, all the potential data in the text and figures were referred to the Saturated Calomel Electrode (SCE). Before the electrochemical impedance spectrum of each replicate was obtained, the sample was subjected to 1 hour of wetting in 0.5 M Na$_2$SO$_4$ solution. Impedance spectra were obtained from a Solartron 1255 FRA coupled to a Solartron EI 1286 and a PC, all controlled by the Zplot software. Impedance spectra collected in the frequency range $10^{-2} \leq f(Hz) \leq 10^6$ were analyzed and interpreted on the basis of equivalent electrical circuits, optimizing the values of the circuit parameters by using Boukamp' program [32].

All the electrochemical experiments were carried out at laboratory temperature (23 ± 2°C) and with the electrochemical cell in a Faraday cage to reduce external interferences as much as possible.

To improve the experimental data reliability, three replicates of each sample type were measured in all the tests.

EXPERIMENTAL RESULTS AND DISCUSSION

Atmospheric Exposure Test

Corrosion of the Metallic Coatings

It is known that all the materials degrade under the influence of atmospheric factors such as oxygen, humidity, and/or pollutants (SO$_2$, NaCl, NO$_x$, etc.). Another important degradation source is the sun radiation, particularly its UV rays. All these influences compose the so-called "Macroclimate" of a determined zone [33]. In change, "Microclimate" is defined as the specific climate formed around an object and it results of vital importance to understand the atmospheric mechanisms causing the materials degradation. Among the parameters used to define it are the surface time of wetness (TOW), the heating by sun radiation, mainly the infrared, and the acidic nature ions (SO_3^{2-}, NO_2^-, Cl^-) gathering within the aqueous layer deposited on the object. On the other hand, the atmospheric corrosion process is the sum of partial corrosion processes taking place each time an electrolyte layer deposits on the surface metal. The rain, snow, fog, and/or humidity condensation produced by temperature changes are the main promoters of atmospheric corrosion. In such sense, the value of some climatological variables characterizing the average exposure conditions corresponding to the station used in the present work are shown in Tables 2 and 3. The aggressiveness of La Plata station was attributed to its high relative humidity, severe and lengthy TOW as well as surface runoff supported by the tested replicates.

Hot-dip zinc is widely used as a coating for carbon steel because of its good corrosion resistance and relatively low price [34]. Due to its practical use, zinc atmospheric corrosion has been studied in field exposures as well as in laboratory with controlled environments [35, 36]. Zincite, ZnO, is the first product formed when the naked metal is exposed to the air, creating a protective film that inhibits corrosion process. Under humidity conditions higher than 80%, zinc is oxidized forming zinc hydroxide. If the pH on the surface is high enough, this hydroxide can react with atmospheric components such as CO_2, SO_x, and Cl^-, forming, in the hydroxide/air interface, the corresponding zinc basic salts [37]. Some of these products form a compact film that protects the metal against later corrosive attacks [38]. An important intermediate in the subsequent formation of other corrosion products is hydrozincite $Zn_4CO_3(OH)_6 \cdot H_2O$ [39, 40]. If the pH of the humid surface is low, neither hydroxide nor basic salts are formed [37]. In presence of SO_2 polluted air, the main corrosion product is hydroxysulfate $Zn_4SO_4(OH)_6 \cdot 4H_2O$, and in presence of Cl^- contamination, the precipitation of insoluble hydroxychloride $Zn_5(OH)_8Cl_2 \cdot H_2O$ is possible.

The hot-dip aluminium-zinc alloy, known as Zincalume, actually contains about 55% aluminium, 1.5% silicon, and the balance zinc. A microstructure of the alloy-coated steel which forms on cooling is essentially two phase, comprising about 80% by volume of a dendritic aluminium-rich phase and the remainder an interdendritic zinc-rich phase with a thin intermetallic layer next to the steel substrate. When the coating corrodes initially, the zinc phase corrodes preferentially until the formation of corrosion products reduces further activity in these areas. During the initial stage of corrosion, the coating behaves like zinc coating. In the later stages of corrosion when the coating is essentially comprised of zinc corrosion products carried in an aluminium-rich matrix, the corrosion becomes more characteristic of the aluminium-rich phase, resulting in a lower corrosion rate, more typical of aluminium [41, 42].

Weight-loss measurements provide the most reliable figure concerning the aggressiveness of a given atmosphere, so that the corresponding corrosion data approach to the service conditions more than any other test. In the present case, as it is shown in Figure 2 both materials presented a linear relationship between its weight-loss and the weathering time and, considering the 12 years exposure, the galvanized coating exhibited a degree of corrosion 4.94 times greater than that of the 55%Al-Zn coating.

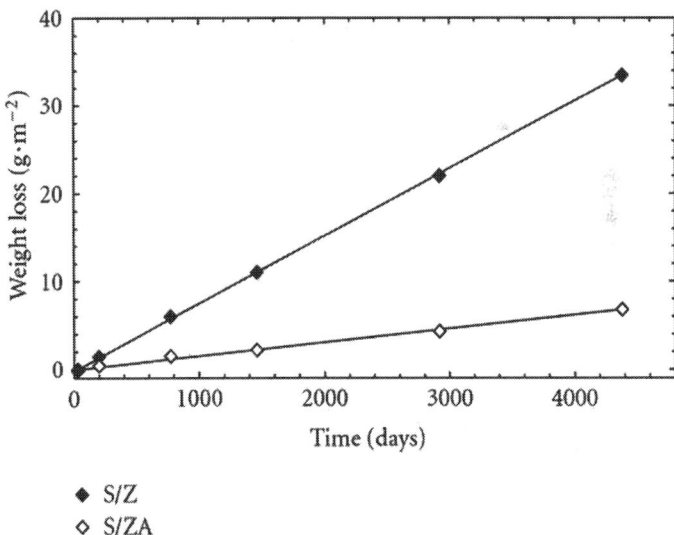

Figure 2: Plots showing the time dependence of the weight loss suffered by the bare S/Z and S/ZA sheets.

In general, the zinc coating suffered uniform corrosion with the development of a layer of corrosion products, mainly ZnO and $Zn_4CO_3(OH)_6 \cdot H_2O$, but in particular places of the surface, the accumulation of atmospheric dust produced localized corrosion as it is possible to see in the SEM images shown in Figure 3. On the other hand, on the 55%Al-Zn coating and due to the complex structure of the alloy, the development of localized corrosion was observed as a consequence of the preferential dissolution of the interdendritic Zn-rich phase, which provoked loss of surface' brightness and the development of thin dark lines related to the aluminium corrosion process.

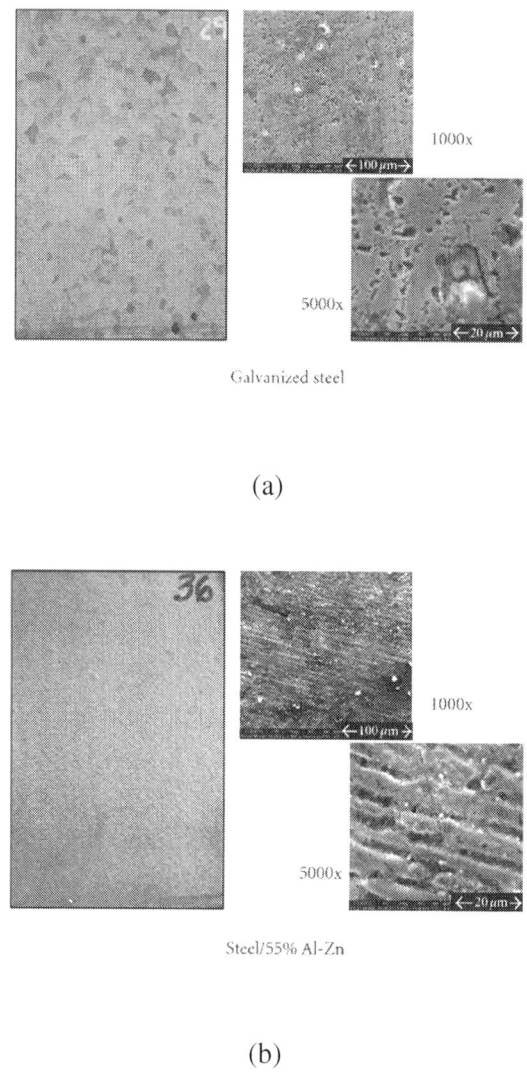

Figure 3: Photographs and SEM images of the bare metallic coatings after 12 years of exposure.

During all the weathering period, both metallic coatings were able to afford cathodic protection to the substrate as it is shown in Figure 4.

Galvanized steel

(a)

Steel/55% Al-Zn

(b)

Figure 4: Photographs of the scribed area of bare metallic coatings after 12 years of exposure.

These results can be explained by considering the climatic conditions prevailing during the outdoor exposure prior to samples removal, that is, level of sulphur compounds, wet/dry cycles, high TOW, and pluvial precipitations dissolving the zinc corrosion products and releasing zinc ions from the corroded surface, which are dispersed to the environment. This phenomenon is known today as a metal runoff process [43–50].

Duplex Systems

As a direct way of evaluating the anticorrosive performance of organic coatings, the exposure test to natural atmospheres either for intact or scribed painted surfaces is, without any doubt, the best.

Experimental results coming from exposure tests to natural weather conditions are further representative of the protective and aesthetic properties provided by the topcoat paint. In such a sense, the results obtained from visual inspections periodically carried out for 12 years of exposure in La Plata station confirmed that weathering of the considered duplex systems proceeds very slowly. The periodical visual inspection put in evidence that no sample presented underrusting, peeling, cracking, or checking. From the 4th year weathering and due to the effect of the UV radiation, the Epoxy paint systems showed significant deterioration by chalking, and the topcoat Alkyd paint began to show significant changes of color, brightness, and chalking that led to the exposure of the primer from the 8th year of test. In any case, blistering or filiform corrosion near the scribed cross was observed. Examples of this behavior are presented on Figures 5 and 6. These results were attributed to the highest resistance of the polyurethane topcoat paint to the UV radiation, rain and temperature changes due to the strong interaction between the reactive components of the polymer and the chemically stable pigment (TiO_2) added to the effective anticorrosive protection offered by the primer.

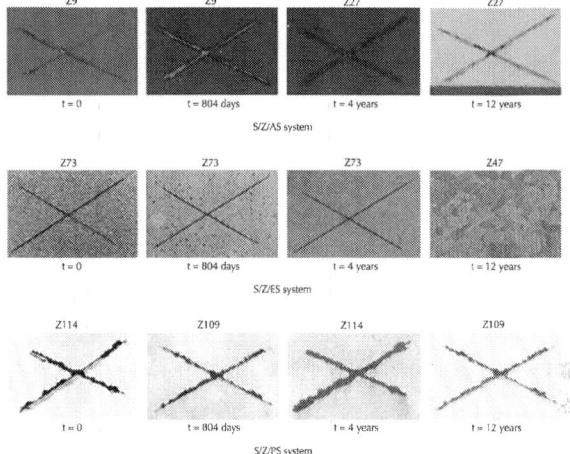

Figure 5: Photographs showing the evolution of S/Z/painting systems as a function of the exposure time.

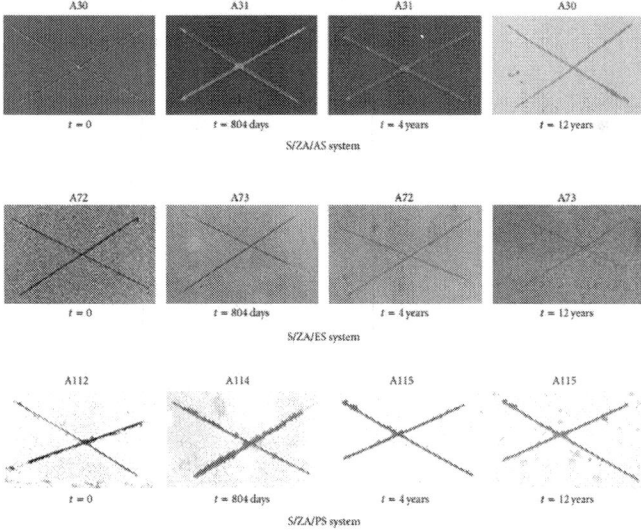

Figure 6: Photographs showing the evolution of S/ZA/painting systems as a function of the exposure time.

At this point, it is noteworthy to remark that besides knowing the behavior of painting systems when they are intact, it is important to evaluate how they performed when mechanical damage occurs. For that purpose, some of the samples were scribed in an X-shape before being exposed to the natural atmosphere.

The main characteristic of duplex systems is to get and maintain good adhesion to the metallic-coated steel surface during its weathering period. As it is shown in Figure 7, independent of the metallic substrate the polyurethane coating presented the largest adhesion loss during the weathering. Due to the previously mentioned behavior of Epoxy and Alkyd paint systems from the 4th year of weathering, is important to mention that in the case of both Alkyd systems the adhesion test considered mainly the primer and in the case of the Epoxy-based samples the evaluation was interrupted as a consequence of the observed chalking degree.

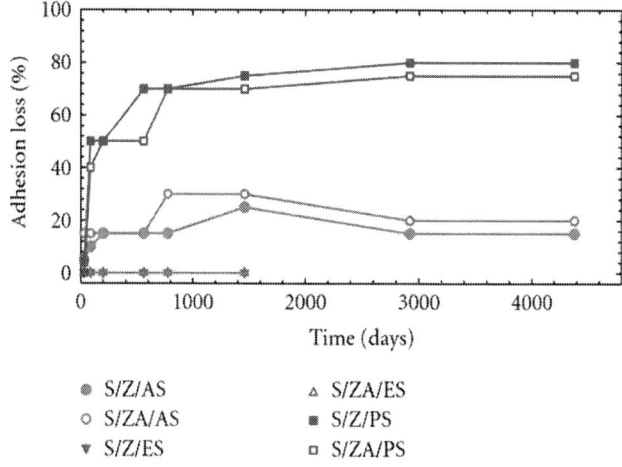

Figure 7: Plots showing the time dependence of the adhesion loss suffered by the painted S/Z and S/ZA sheets.

Electrochemical Tests

During atmospheric corrosion, in general, the metal is not immersed in large quantities of electrolyte but in contact with thin layers or monolayers of moisture, due to that the corrosion process develops as localized corrosion cells. In that situation, the measurement of the corrosion potential as well as of the resistive and capacitive parameters governing the electrochemical behavior of the metal/coating interface is not always possible during the atmospheric corrosion [51, 52].

According to Zhang and Lyon [52], the cathodic process for metals like steel, zinc, and copper coated with thin (<100 µm) water films reveals a diffusional limiting current whose value depends on the water film thickness. For thinner thicknesses, like in most of the atmospheric corrosion cases, the main cathodic process is controlled by activation. In the case of zinc, due to its high electronegativity, the cathodic process is not sensitive to the water film thickness present on the surface. On the other hand, as it was mentioned earlier, the protective capacity of the corrosion products formed during the atmospheric exposure depends on a variety of properties, which are also dependent on the composition and metallurgical history of the metal as well as on the atmospheric variables [1, 5, and 53].

Corrosion Potential and Impedance Results for Bare S/Z and S/ZA Sheets

The measured corrosion potential values (E_{corr}) point out the metal susceptibility to be corroded. In general, when a value for a given medium is nobler (positive), it will result more resistant to corrosion. As it is possible to see in Figure 8, the corrosion potential evolution for the bare coating materials (Z and ZA) along the 12 years of exposure was quite similar. In this case, the surface of both the bare S/Z and the S/ZA sheets remained active up to the end of the test exposure with potential values ranging between −1.04 V/SCE and −1.00 V/SCE, which are characteristics of these metals under free corrosion processes. [54] The continuous electrochemical reactivity was attributed to the rain runoff effect on the corrosion products, which avoided the formation of an oxide, hydroxide, and/or passive protective layer on the bare surfaces.

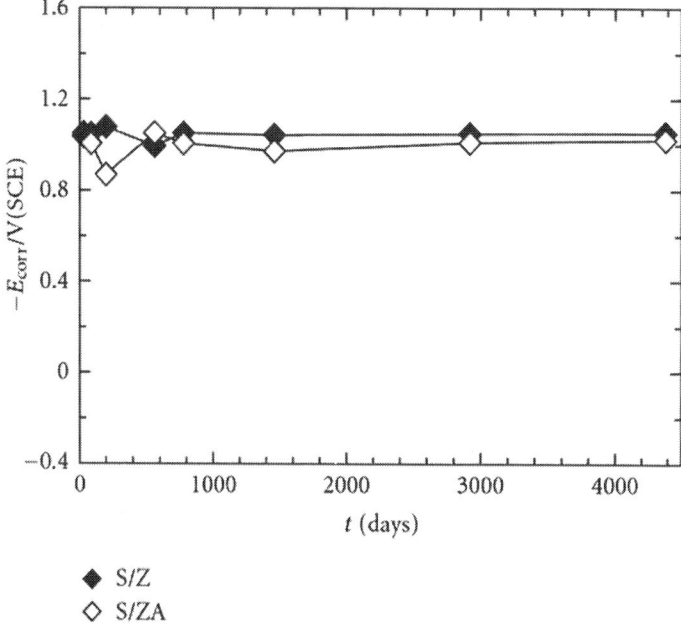

◆ S/Z
◇ S/ZA

Figure 8: Plots showing the time dependence of the E_{corr} values of bare S/Z and S/ZA sheets as a function of the exposure time.

These data are in accordance not only with the visual inspection but also with the results obtained elsewhere [55] by optic and electronic microscopy techniques, which put in evidence a developed corrosion process. The surface had hollows and corrosion products characterized as oxides and basic carbonates. On the other hand, the 55%Al-Zn alloy had a big cathodic area where the metals exposed to the atmosphere will corrode by coupling with the oxygen cathodic reduction reaction:

$$O_2 + 2H_2O + 4e^- \longrightarrow 4OH^- \qquad (1)$$

Although when the level of contamination with acid products is high, the hydrogen evolution gets importance as cathodic reaction:

$$2H^+ + 2e^- \longrightarrow H_2 \qquad (2)$$

Regardless of which reaction prevails, the pH on the cathodic region increases. From a certain level of acidity, it is possible that the SO_2 of a polluted atmosphere acts as an oxidant able to impart a great acceleration to the cathodic process.

The impedance spectra of the coating Zn (Figure 9) may be interpreted in terms of the corrosion products film structure that is usually formed on the surface. The first time constant (R_1C_1) may be linked with the compact inner layer of ZnO and the second one (R_2C_2) with the external and porous layer of $Zn_4CO_3(OH)_6 \cdot H_2O$ [55,56]. This surface film seems to inhibit further metal dissolution, although the environmental conditions determine the extent of corrosion progress due to a competition between film formation and film removal reactions. It was found that data of zinc corrosion measurements correlate with the air pollution levels given as a function of the SO_2 and Cl^- concentrations [57].

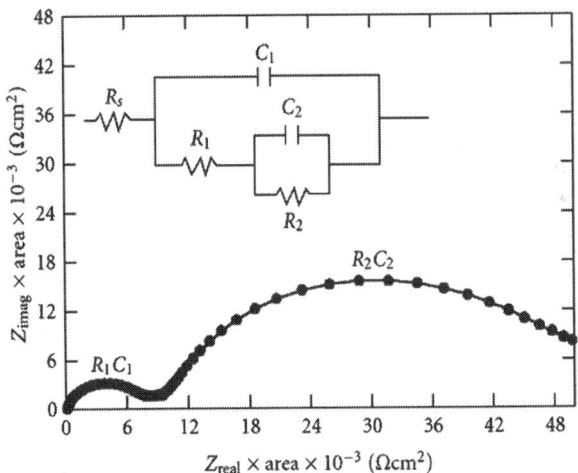

Figure 9: Equivalent circuit model used for fitting the tested bare S/Z and S/ZA sheets.

For similar exposure conditions, the influence of the coating composition on the bare sheet impedance values is shown in Figure 10. In it can be seen that the charge transfer resistance (R_1) values shown by the 55%Al-Zn alloy was slightly higher than that of the zinc layer but the resistive contribution (R_2) of the external and porous layer to the system total impedance was very similar. These results are in accordance with the electrochemical activation demonstrated by both metallic surfaces.

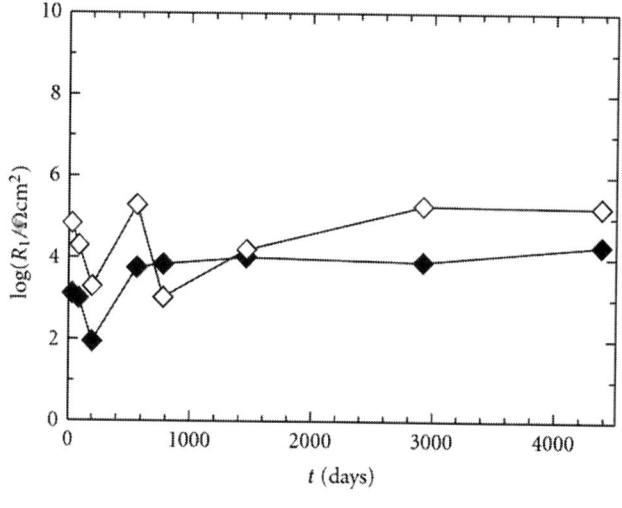

(a)

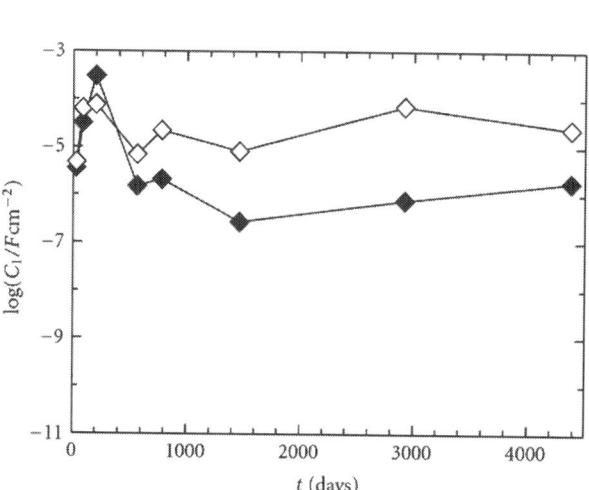

(b)

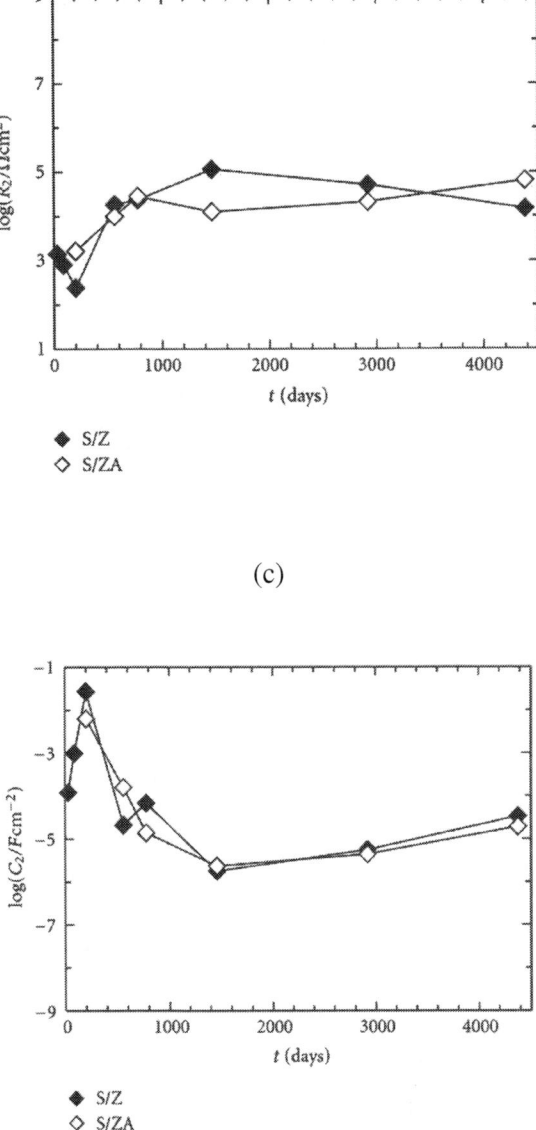

(c)

(d)

Figure 10: Evolution of log R_1, log C_1, log R_2, and log C_2 parameters of the tested bare S/Z and S/ZA sheets.

Corrosion Potential and Impedance Results for Painted S/Z and S/ZA Sheets

Rest or corrosion potential (E_{corr}) measurements for painted metals and their time dependence have been questioned with regard to their use as a technique for evaluating the anticorrosive resistance of organic coatings [58]. However, its changes as a function of the exposure time to aqueous media have been successfully used as a simple tool to study the corrosion protection afforded by organic coatings [59–62]. Depending upon the microstructure of the paint coating, especially its polymerization degree, a certain period elapses until electrolyte penetration channels are established through which the underlying metal comes into contact with the medium. So, it is not surprising that, when a compact structure and high crosslinking level are accompanied by an also high film thickness, a few days of testing is not enough time for the electrolyte to enter in contact with the base metal of coated specimens, form the electrochemical double layer, and enable the measurement of a corrosion potential.

Figure 11 shows the corrosion potential (E_{corr}) values measured for each coated steel sheet exposed to the natural atmosphere of La Plata station. As can be seen, the (E_{corr}) values measured almost from the beginning and up to the end of the test for S/Z/AS, S/ZA/AS, S/Z/ES, and S/ZA/ES were quite similar to those obtained for the bare S/Z and S/ZA sheets (between −0.9 and −1.1 V/SCE). This means that, at least from the thermodynamic point of view, the protective properties offered by the alkyd- and epoxy-based painting systems were not sufficiently effective as to avoid the onset of the underlying zinc or 55%Al-Zn corrosion. On the other hand, the polyurethane-based painting system offered much more promising protective properties, particularly when applied on S/Z sheets, since the S/Z/PS system potential values remained in an electrochemically passive zone. A similar performance was supplied by the S/ZA/PS up to almost reaching the 12 years of exposure where together with the measurement of an $E_{corr} \approx -1.00$ V/SCE, the first sign of a localized corrosion was detected by EIS.

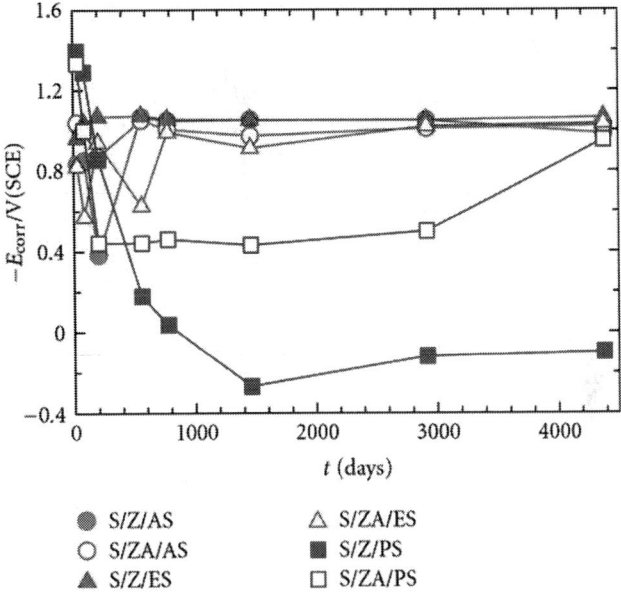

Figure 11: Plots showing the time dependence of the E_{corr} values of painted S/Z and S/ZA sheets as a function of the exposure time.

Since the main difference among the S/Z/painting systems and S/ZA/painting systems was the applied paint formulation used in each case, it is assumed that the magnitude of the E_{corr} displacements may be particularly associated with both the relative easiness with which the climatic variables affect the paint film structure and, hence, its protective properties. However, and it will be discussed in the next paragraph, except in the case of the epoxy-based painting system, the other two were able to protect relatively (AS) and effectively (PS) the metallic substrate from the corrosive atmosphere. This conclusion arises from the fact that the corrosion potential values measured for the S/Z/PS and S/ZA/PS panels were mostly nobler than the corresponding to bare S/Z or S/ZA sheets subjected to the same experimental conditions. This effective protection was mainly attributed to the PS barrier painting system, which could resist the strong aggressive action coming from the atmospheric conditions.

The impedance modulus ($|Z|$) of replicated samples as a function of their exposure time to the natural atmosphere of La Plata station illustrates Figure 12. A fast and simple qualitative analysis of this figure allows to infer that both the shape of all the experimental diagrams was fairly similar and it is possible to presume the presence of at least two time constants, one at low frequencies and another at high frequencies.

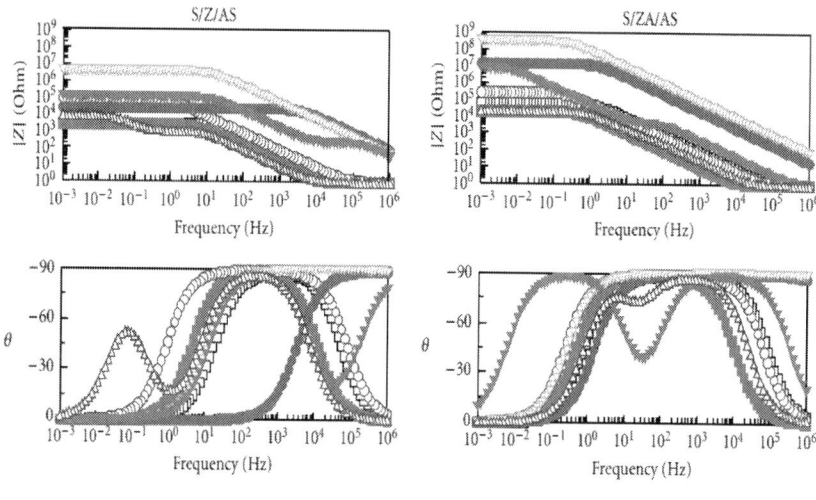

(a)

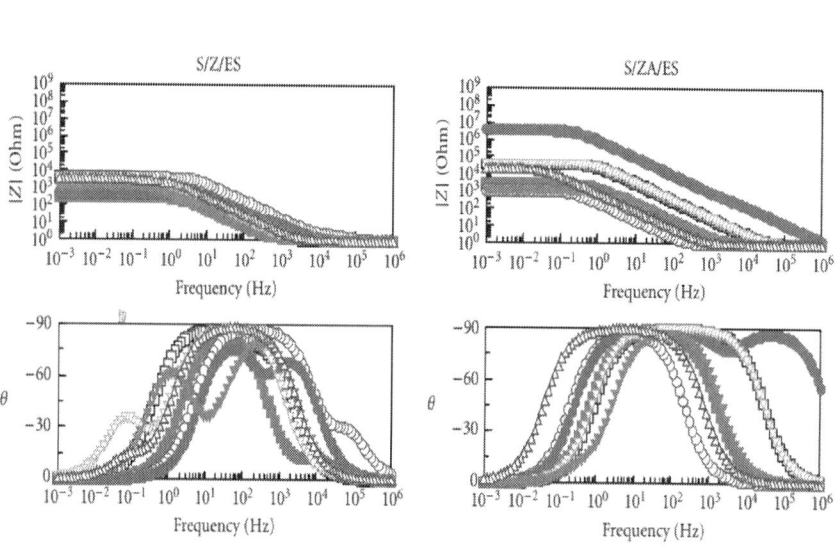

(b)

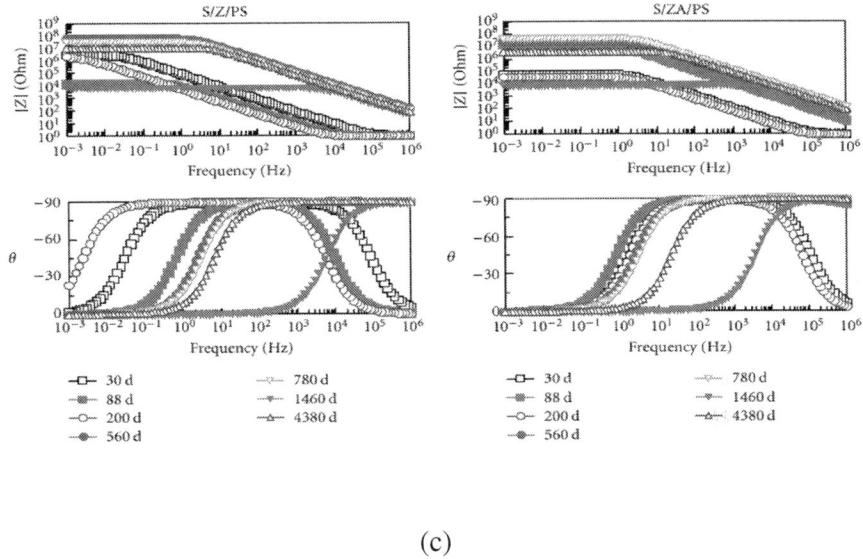

(c)

Figure 12: Bode plots showing the time dependence of the duplex systems impedance during their exposure to the natural atmosphere of La Plata station for 12 years.

As seen in Figure 12, all the tested systems showed changes more or less significant of their (|Z|) and phase angle (Theta) during the weathering period. The fluctuating impedance values can be attributed to the dynamic behavior of the painting system structure frequently subjected to wet/dry cycles and/or other climatic changes as well as of the metal/paint interface through which the corrosion products gathered at the bottom and/or within the coating defects enhanced the coating barrier protection and, therefore, contribute to an increase in the impedance of the protective system at medium and low frequencies; however, as the time elapses, new defects appear at the weaker (less protective areas) paint layer allowing the inducing corrosion species permeation, and, consequently, the development of new electrochemically active zones.

The fact that the initial substrate attack is localized could be ascribed to the presence of very small defects in the paint layer, which act as an electrical shunt. As the exposure time goes on, the equilibrium between the development rate of the corrosion products and their diffusion rate towards the outdoor medium may be reached and, consequently, the total impedance fluctuations become small.

Equivalent Circuit Models

The painting system as well as the S/Z or S/ZA substrates deterioration takes place from processes having a complex nature. Consequently, to interpret and explain in electrochemical terms the time dependence of the acquired impedance data, it has been necessary to propose appropriate equivalent circuit models.

Impedance spectra provide useful information concerning the evolution of both the protective features of the organic coating and the kinetics of the underlying metallic substrate corrosion process as a function of the exposure time to experimental or real service conditions. Thus, the dynamic character of the painting system barrier properties, the anticorrosive action of specific pigments, the corrosion products formation, and also changes in the disbonded area are accounted for the time dependence of the coated steel/medium impedance spectra. In general, an explanation of why and how such changes take place can be given by associating them to the resistive and capacitive parameters derived from fitting impedance data with nonlinear least squares algorithms involving the transfer function of the equivalent circuit model shown in Figure 13, [63–67]. They represent the parallel and/or series connection of some resistors and capacitors, simulating a heterogeneous arrangement of electrolytically conducting paths, where R represents the electrolyte resistance between the reference and working (coated steel) electrodes, R_c (resistance to the ionic flux) describes paths (pores, low crosslinking) of lower resistance to the electrolyte diffusion short-circuiting the paint film, and C_c is the dielectric capacitance representing the intact part of the same paint film [68]. Once the permeating and corrosion-inducing chemicals (water, oxygen and ionic species) reach electrochemically active areas of the substrate, particularly at the bottom of the paint film defects, the metallic corrosion become to be measurable so that it's associated parameters, the electrochemical double-layer capacitance, C_{dl}, and the charge transfer resistance, R_t, can be estimated. It is important to remark that the values of these parameters vary direct (C_{dl}) and inversely (R_t) with the size of the corroding area.

On the other hand, distortions observed in these resistive-capacitive contributions indicate a deviation from the theoretical models in terms of a time constants distribution due to either lateral penetration of the electrolyte at the steel/paint interface (usually started at the base of intrinsic or artificial coating defects), underlying metallic substrate surface heterogeneity (topological, chemical composition, surface energy), and/or diffusion processes that could take place along the test [69, 70]. Since all these factors make the impedance/frequency relationship nonlinear, they are taken into consideration by replacing one or more capacitive components (C_i) of the equivalent circuit transfer function by the corresponding constant phase element Q_i (CPE), whose impedance dispersion relation is given by $Z = (j\omega)^{-n}/Y_0$ and $n =$ CPE power $= a/(\pi/2)$ [32, 71].

Difficulties in providing an accurate physical description of the occurred processes are sometimes found. In such cases, a standard deviation value ($\chi^2 < 5 \times 10^{-4}$) between experimental and fitted impedance data may be used as final criterion to define the most probable circuit.

According to the impedance data dispersion, the fitting process was performed using either the dielectric capacitance C_i or the phase constant element Q_i; however, the C_i parameter was used in the following plots to facilitate the results visualization and interpretation.

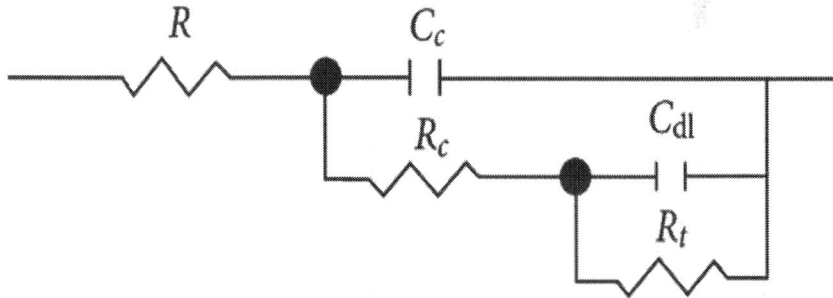

Figure 13: Equivalent circuit model used for fitting the tested duplex systems.

Time Dependence of the Impedance Resistive and Capacitive Components

The values of the resistive and capacitive components of the impedance corresponding to all the painted samples exposed to the natural atmosphere of La Plata station for 12 years are shown in Figure

14. As seen, the S/Z/PS and S/ZA/PS samples offered an excellent anticorrosive performance up to the end of the exposure. This behavior could be attributed to its excellent barrier effect due to the structurally homogeneous and strong paint film ($R_c \approx 10^7$–10^8 Ωcm^2, $C_c \approx 10^{-10}$–10^{-9} Fcm^{-2}), which was able to counteract the significant adhesion loss suffered by this painting system during weathering and slowed down the development of the alloy coating corrosion process up to the end of the exposure.

For the other two sample types, a rather highly fluctuating R_c values (two or more orders of magnitude) were found within the first 1400 days of exposure but then, and up to the end of the test, they remained changing between 106–104 $\Omega cm2$; on the other hand, its coupled dielectric capacitance (C_c) followed the same unstable trend at the beginning of the test but then, due to the deterioration degree reached by the AS and ES painting systems as a consequence of the adverse climatic conditions, led to C_c values ($\approx 10^{-6}$–10^{-5} Fcm^{-2}), that is, close to the bare S/Z and S/ZA sheets.

On the other hand, the same Figure shows great differences in the electrochemical response ($R_t C_{dl}$) of the different systems. In the case of the polyurethane systems, the corrosion process was either absent (S/Z/PS) or at least its development was delayed up to the end of exposure (S/ZA/PS). On the contrary, for the rest of the considered duplex systems after variable induction periods, the metallic coating degradation was detected. The worst corrosion protection was afforded by the Epoxy-based painting system since the corrosion process was detected by EIS at 30 (S/Z/ES) and 200 (S/ZA/ES) days. This behavior would be ascribed to the different electrochemical reactivity of the Z and ZA coatings. In the case of the Alkyd system, the induction period was of 600 and 1400 days for S/Z/As and S/ZA/AS, respectively. Again, the better performance of the last system could be accredited to a relatively good, although too short for practice purposes, barrier effect offered by the paint film added to the lower electrochemical reactivity of the S/ZA sheet.

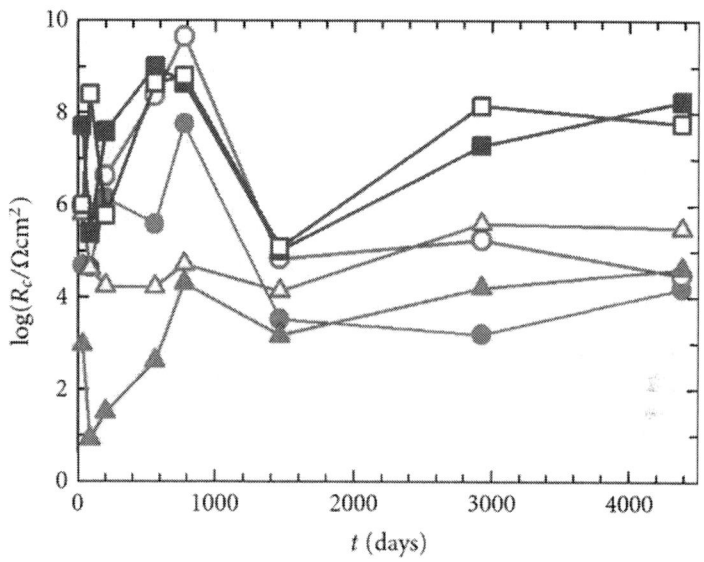

(a)

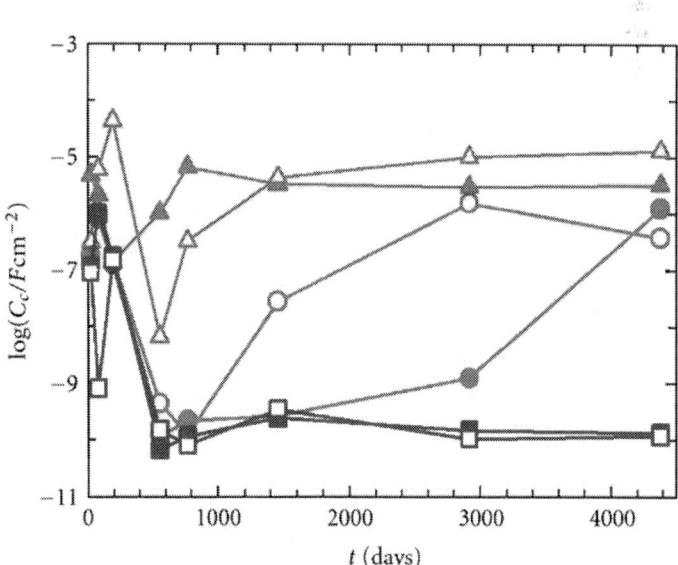

(b)

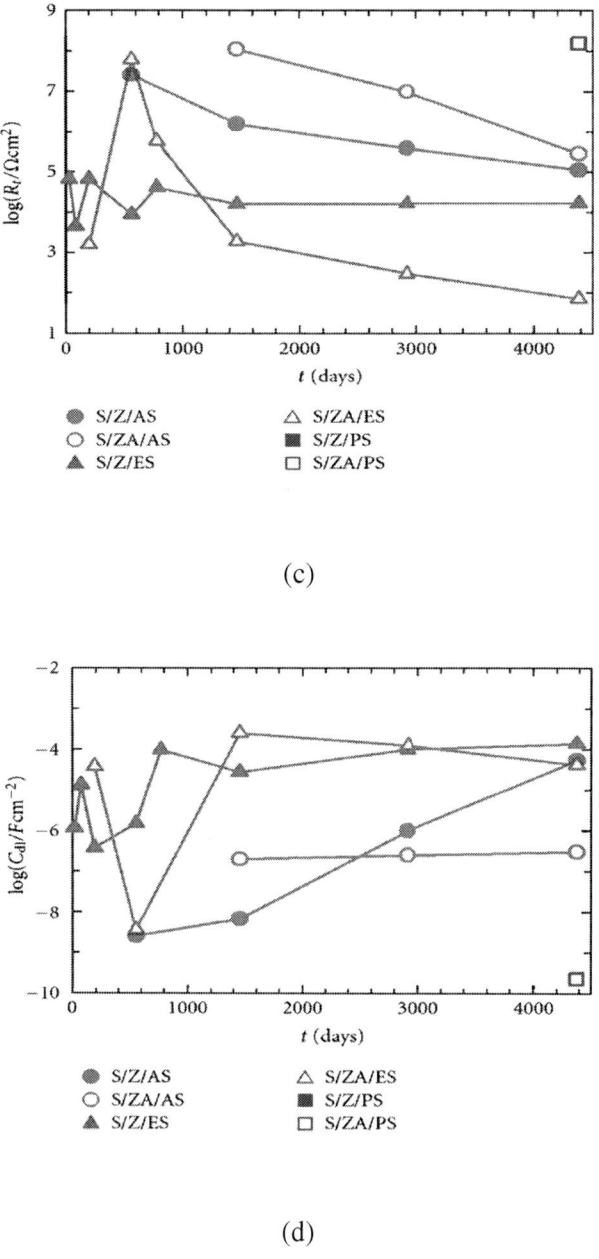

Figure 14: Evolution of R_c, C_c, R_t and C_{dl} parameters of the duplex systems impedance during their exposure to the natural atmosphere of La Plata station for 12 years.

CONCLUSIONS

At the end of this work, it is possible to summarize some conclusions valid for the studied materials.

All the laboratory and field tests involved in this work were useful to understand the behavior of the studied duplex systems subjected to natural weathering at La Plata Station. The good correlation between visual inspection and electrochemical tests allowed explaining some troubles observed in practice and, on this base, contribute to solve them to maintain its useful life as long as possible.

An almost constant corrosion rate of bare zinc and zinc-aluminum layers acting as galvanic coating of steel sheets was found during the long-term exposure to the natural atmosphere of the La Plata station. Both materials were able to cathodically protect the steel substrate for 12 years.

Regarding the comparative study among the three painting systems applied on S/Z or S/ZA sheets, different $R_c C_c$ and $R_t C_{dl}$ evolutions were obtained depending mainly on the paint. The best protective performance offered by the Polyurethane-based painting system was explained in principle taking in account its better barrier properties. The experimental results coming from the alkyd- and epoxy-based painting systems were not satisfactory due to their low resistance to the atmospheric conditions existing at La Plata station.

Despite the interface degradation (loss of adhesion) shown by all the painting systems, the corrosion process did not progress from the cross cut towards the underlying metallic substrate.

ACKNOWLEDGMENTS

The authors thank the Comisión de Investigaciones Científicas de la Provincia de Buenos Aires (CICPBA), the Consejo Nacional de Investigaciones Científicas y Técnicas (CONICET), and the Universidad Nacional de La Plata for the financial support to carry out the present paper.

REFERENCES

1. M. Stratmann, K. Bohnenkamp, and W. J. Engell, "An electrochemical study of phase-transitions in rust layers," Corrosion Science, vol. 23, no. 9, pp. 969–985, 1983.
2. L. Ferretti, E. Traverso, and G. Ventura, "Marine corrosion of mild steel in a thermically altered natural environment," Anti-Corrosion Methods and Materials, vol. 23, no. 5, pp. 3–5, 1976.
3. C. E. Bird and F. J. Strauss, "Effect of wet storage staining on subsequent atmospheric corrosion of galvanized iron sheets," Materials Performance, vol. 15, no. 11, p. 27, 1976.
4. F. E. Goodwin, "Mechanism of corrosion of Zinc and Zinc -5% Aluminum steel sheet coatings," in Zinc-Based Steel Coatings Systems: Metallurgy and Performance, G. Krauss, Ed., pp. 183–193, The Minerals, Metals & Materials Society, Warrendale, Pa, USA, 1990.
5. M. Pourbaix, "Une méthode éléctrochimique rapide de predetermination de la corrosion atmospherique," Tech. Rep. 1, CEBELCOR, 1969.
6. Y. Suzuki, Y. Hisamatsu, and N. Masuko, "Nature of atmospheric rust on iron," Journal of the Electrochemical Society, vol. 127, no. 10, pp. 2210–2214, 1980.
7. ASM Handbook, vol. 13 of Corrosion: Materials, ASM International, 1992.
8. J. C. Zoccola, H. E. Townsend, A. R. Borzillo, and J. B. Horton, "Atmospheric factors affecting the corrosion of engineering metals," in Proceedings of the American Society for Testing and Materials (STP ‹78), vol. 646, pp. 165–184, 1978.
9. F. C. Porter, Zinc Handbook: Properties, Processing and Use in Design, Marcel Dekker, New York, NY, USA, 1991.
10. J. F. H. van Eijnsbergen, Duplex Systems, vol. 7, Elsevier, Amsterdam, The Netherlands, 1994.
11. K. L. Mittal, "Stresses and adhesion of thin metallic coatings on oxide substrates," in Adhesion Measurements of Film and Coatings, K. L. Mittal, Ed., pp. 1–13, VSP, Utrecht, The Netherlands, 1995.

12. Steel Structure Painting Manual. Systems and Specifications, vol. 2, Steel Structures Painting Council, Pittsburgh, Pa, USA, 7th edition, 1995.
13. J. F. Malone, "Painting hot dip galvanized steel," Materials Performance, vol. 31, no. 5, pp. 39–42, 1992.
14. C. H. Hare, "Corrosion and the preparation of metals for painting," in Unit 26 Federation Series on Coatings Technology, Federation of Societies for Coatings Technology, Philadelphia, Pa, USA, 1978.
15. K. W. Allen, Strength and Structures. Aspect of Adhesion, vol. 1, University Press of London, London, UK, 1965.
16. T. R. Bullet and A. T. S. Rudram, "The coating and the substrate," Journal of the Oil and Colour Chemists› Association, vol. 44, pp. 787–807, 1961.
17. C. H. Hare, Good Painting Practice. Steel Structure Painting Manual, Steel Structures Painting Council, Pittsburgh, Pa, USA, 3rd edition, 1995.
18. R. Barnhart, D. Mericle, Ch. Mobley, T. Hocking, J. H. Bogran, and E. McDaniel, "Why surface preparation is important?" Journal of Protective Coatings and Linings, vol. 14, no. 9, pp. 61–64, 1997.
19. H. Leidheiser Jr., "Corrosion of painted metals—a review," Corrosion, vol. 38, no. 7, pp. 374–383, 1982.
20. I. Sekine, M. Yuasa, N. Hirose, and T. Tanaki, "Degradation evaluation of corrosion protective coatings by electrochemical, physicochemical and physical measurements," Progress in Organic Coatings, vol. 45, no. 1, pp. 1–13, 2002.
21. G. Rocchini, "The importance of choosing the correct electrochemical technique for evaluating corrosion rates," Corrosion Prevention and Control, vol. 48, no. 4, pp. 125–134, 2001.
22. S. Maeda, "Surface chemistry of galvanized steel sheets relevant to adhesion performance," Progress in Organic Coatings, vol. 28, no. 4, pp. 227–238, 1996.
23. R. L. Howard, I. M. Zin, J. D. Scantlebury, and S. B. Lyon, "Inhibition of cut edge corrosion of coil-coated architectural cladding," Progress in Organic Coatings, vol. 37, no. 1, pp. 83–90, 1999.

24. R. L. Howard, S. B. Lyon, and J. D. Scantlebury, "Accelerated tests for the prediction of cut-edge corrosion of coil-coated architectural cladding. Part I: cyclic cabinet salt spray," Progress in Organic Coatings, vol. 37, no. 1, pp. 91–98, 1999.
25. R. L. Howard, S. B. Lyon, and J. D. Scantlebury, "Accelerated tests for prediction of cut edge corrosion of coil-coated architectural cladding. Part II: cyclic immersion," Progress in Organic Coatings, vol. 37, no. 1, pp. 99–105, 1999.
26. S. Feliu, V. Barranco, and S. Feliu, "Contradictory results of the UVCON and saline immersion tests regarding the evaluation of some inhibitor/lacquer combinations on galvanised coatings," Progress in Organic Coatings, vol. 50, no. 3, pp. 199–206, 2004.
27. A. R. di Sarli and R. A. Armas, "An assessment of the anti-corrosive properties of epoxy paints. Correlation between impedance measurements and the salt-spray cabinet test," Corrosion Prevention & Control, vol. 36, no. 5, pp. 127–131, 1989.
28. P. R. Seré, J. D. Culcasi, C. I. Elsner, and A. R. Di Sarli, "Study of the corrosion process at the galvanized steel/organic coating interface," in Proceedings of the SCANNING ‹98, vol. 20, no. 3, pp. 274–275, Baltimore, Md, USA, 1998.
29. B. M. Rosales, A. R. di Sarli, F. Fragata et al., "PATINA network—performance of coil coating in natural atmospheres of Ibero-America," Revista de Metalurgia, vol. Extra, pp. 201–205, 2003.
30. B. M. Rosales, A. R. di Sarli, O. de Rincón, A. Rincón, C. I. Elsner, and B. Marchisio, "An evaluation of coil coating formulations in marine environments," Progress in Organic Coatings, vol. 50, no. 2, pp. 105–114, and 2004.
31. B. Del Amo, L. Véleva, C. I. Elsner, and A. R. di Sarli, "Performance of coated steel systems exposed to different media: part I. Painted galvanized steel," Progress in Organic Coatings, vol. 50, no. 3, pp. 179–192, 2004.
32. B. A. Boukamp, Report CT88/265/128, CT89/214/128, University of Twente, Amsterdam, the Netherlands, 1989.
33. E. V. Schmid, Exterior Durability of Organic Coatings, FMJ International Publications, Redhill, UK, 1988.
34. S. C. Chung, A. S. Lin, J. R. Chang, and H. C. Shih, "EXAFS study of atmospheric corrosion products on zinc at the initial stage," Corrosion Science, vol. 42, no. 9, pp. 1599–1610, 2000.

35. Q. Qu, C. Yan, Y. Wan, and C. Cao, "Effects of NaCl and SO_2 on the initial atmospheric corrosion of zinc," Corrosion Science, vol. 44, no. 12, pp. 2789–2803, 2002.
36. J. Morales, S. Martín-Krijer, F. Díaz, J. Hernández-Borges, and S. González, "Atmospheric corrosion in subtropical areas: influences of time of wetness and deficiency of the ISO 9223 norm," Corrosion Science, vol. 47, no. 8, pp. 2005–2019, 2005.
37. V. Kucera and E. Mattsson, Corrosion Mechanisms, Marcel Dekker, New York, NY, USA, 1987.
38. T. E. Graedel, "Corrosion mechanisms for zinc exposed to the atmosphere," Journal of the Electrochemical Society, vol. 136, no. 4, pp. 193C–203C, 1989.
39. X. G. Zhang, Corrosion and Electrochemistry of Zinc, Plenum Press, New York, NY, USA, 1996.
40. F. Mansfeld, Corrosion Mechanisms, Marcel Dekker, New York, NY, USA, 1987.
41. H. E. Townsend, L. Allegra, R. J. Dutton, and S. A. Kriner, "Hot-dip coated sheet steels—a review," Materials Performance, vol. 25, no. 8, pp. 36–46, 1986.
42. H. E. Townsend and A. R. Borzillo, "Twenty-year atmospheric corrosion tests of hot-dip coated sheet steel," Materials Performance, vol. 26, no. 7, pp. 37–41, 1987.
43. J. W. Spence, F. H. Haynie, F. W. Lipfert, S. D. Cramer, and L. G. McDonald, "Atmospheric corrosion model for galvanized steel structures," Corrosion, vol. 48, no. 12, pp. 1009–1019, 1992.
44. I. W. Odnevall Wallinder, P. Verbiest, W. He, and C. Leygraf, "Effects of exposure direction and inclination of the runoff rates of zinc and copper roofs," Corrosion Science, vol. 42, no. 8, pp. 1471–1487, 2000.
45. S. Bertling, I. Odnevall, C. Leygraf, and D. Berggren, "Environmental effects of zinc runoff from roofing materials- A new multidisciplinary approach," in Outdoor Atmospheric Corrosion, ASTM STP 1421, H. E. Townsend, Ed., pp. 200–215, American Society for Testing and Materials International, West Conshohocken, Pa, USA, 2002.
46. W. He, I. Odnevall, and C. Leygraf, "Runoff rates of zinc—a four-year field and laboratory study," inOutdoor Atmospheric

Corrosion, ASTM STP 1421, H. E. Townsend, Ed., pp. 216–229, American Society for Testing and Materials International, West Conshohocken, Pa, USA, 2002.

47. J. H. Sullivan and D. A. Worsley, "Zinc runoff from galvanised steel materials exposed in industrial/marine environment," British Corrosion Journal, vol. 39, no. 4, pp. 282–288, 2002.

48. S. Matthes, S. Cramer, S. Bullard, B. Covino, and G. Holcomb, "Atmospheric corrosion and precipitation runoff from zinc and zinc alloy surfaces," in Proceedings of the 58th Annual Conference Corrosion, NACE, 2003, Paper 03598.

49. S. Jouen, B. Hannoyer, A. Barbier, J. Kasperek, and M. Jean, "A comparison of runoff rates between Cu, Ni, Sn and Zn in the first steps of exposition in a French industrial atmosphere," Materials Chemistry and Physics, vol. 85, no. 1, pp. 73–78, 2004.

50. D. Reiss, B. Rihm, C. Thöni, and M. Faller, "Mapping stock at risk and release of zinc and copper in Switzerland-dose response functions for runoff rates derived from corrosion rate data," Water, Air, and Soil Pollution, vol. 159, no. 1, pp. 101–113, 2004.

51. R. M. Kain and E. A. Baker, ASTM STP 947, ASTM, Philadelphia, Pa, USA, 1986.

52. S. H. Zhang and S. B. Lyon, "The electrochemistry of iron, zinc and copper in thin layer electrolytes,"Corrosion Science, vol. 35, no. 1–4, pp. 713–718, 1993.

53. A. Meneses Garcia, J. Rodarte Corro, O. Vidal Gutierrez, and E. Gonzalez Guzman, "The Mexican National Cancer Institute: case report," Revista Del Instituto Nacional de Cancerologia, vol. 36, no. 4, pp. 1179–1182, 1990.

54. M. Pourbaix, Atlas of Electrochemical Equilibria in Aqueous Solutions, NACE, Houston, Tex, USA, 1974.

55. E. A. Sacco, J. D. Culcasi, C. I. Elsner, and A. R. di Sarli, "Evaluation of the protective performance of several duplex systems exposed to industrial atmosphere," Latin American Applied Research, vol. 32, no. 4, pp. 307–311, 2002.

56. J. M. Costa and M. Vilarrasa, "Effect of air pollution on atmospheric corrosion of zinc," British Corrosion Journal, vol. 28, no. 2, pp. 117–120, 1993.

57. M. W. Kendig, A. T. Allen, and F. Mansfeld, "Optimized collection of ac impedance data," Journal of the Electrochemical Society, vol. 131, no. 4, pp. 935–936, 1984.
58. J. Wolstenholme, "Electrochemical methods of assessing the corrosion of painted metals-a review,"Corrosion Science, vol. 13, no. 7, pp. 521–530, 1973.
59. R. A. Armas, C. Gervasi, A. R. di Sarli, S. G. Real, and J. R. Vilche, "Zinc-rich paints on steels in artificial seawater by electrochemical impedance spectroscopy," Corrosion, vol. 48, no. 5, pp. 379–383, 1992.
60. J. E. O. Mayne, "The mechanism of the inhibition of the corrosion of iron and steel by means of paint,"Official Digest, vol. 24, pp. 127–136, 1952.
61. L. Meszáros and S. A. Lindquist, "Study of the performance of zinc-rich paints coatings," in Proceedings of the 115th Event of the European Federation of Corrosion (EUROCORR '82), pp. 147–156, Budapest, Hungary, October 1982, Section 11.
62. M. Morcillo, R. Barajas, S. Feliu, and J. M. Bastidas, "A SEM study on the galvanic protection of zinc-rich paints," Journal of Materials Science, vol. 25, no. 5, pp. 2441–2446, 1990.
63. B. del Amo, L. Véleva, A. R. di Sarli, and C. I. Elsner, "Performance of coated steel systems exposed to different media: part I. Painted galvanized steel," Progress in Organic Coatings, vol. 50, no. 3, pp. 179–192, 2004.
64. O. Ferraz, E. Cavalcanti, and A. R. di Sarli, "The characterization of protective properties for some naval steel/polimeric coating/3% NaCl solution systems by EIS and visual assessment," Corrosion Science, vol. 37, no. 8, pp. 1267–1280, 1995.
65. P. R. Seré, D. M. Santágata, C. I. Elsner, and A. R. di Sarli, "The influence of the method of application of the paint on the corrosion of the substrate as assessed by ASTM and electrochemical method," Surface Coatings International, vol. 81, no. 3, pp. 128–134, 1998.
66. D. M. Santágata, P. R. Seré, C. I. Elsner, and A. R. di Sarli, "Evaluation of the surface treatment effect on the corrosion performance of paint coated carbon steel," Progress in Organic Coatings, vol. 33, no. 1, pp. 44–54, 1998.

67. P. R. Seré, A. R. Armas, C. I. Elsner, and A. R. di Sarli, "The surface condition effect on adhesion and corrosion resistance of carbon steel/chlorinated rubber/artificial sea water systems," Corrosion Science, vol. 38, no. 6, pp. 853–866, 1996.
68. H. Leidheiser Jr. and M. W. Kendig, "Mechanism of corrosion of polybutadiene-coated steel in aerated sodium chloride," Corrosion, vol. 32, no. 2, pp. 69–75, 1976.
69. T. Szauer and A. Brandt, "Impedance measurements on zinc-rich paints," Journal of the Oil and Colour Chemists' Association, vol. 67, pp. 13–17, 1984.
70. D. J. Frydrych, G. C. Farrington, and H. E. Townsend, "The barrier properties of thin carbonaceous films formed by ion beam assisted deposition," in Corrosion Protection by Organic Coatings, M. W. Kendig and H. Leidheiser Jr., Eds., vol. 87-2, p. 240, The Electrochemical Society, Pennington, NJ, USA, 1987.
71. E. P. M. van Westing, G. M. Ferrari, F. M. Geenen, and J. H. W. van de Wit, "In situ determination of the loss of adhesion of barrier epoxy coatings using electrochemical impedance spectroscopy," Progress in Organic Coatings, vol. 23, no. 1, pp. 89–103, and 1993.

Chapter 4

The Addition of Graphene to Polymer Coatings for Improved Weathering

Nurxat Nuraje[1], Shifath I. Khan[2], Heath Misak[2], and Ramazan Asmatulu[2]

[1]Department of Materials Science and Engineering, Massachusetts Institute of Technology, 77 Massachusetts Avenue, Cambridge, MA 02139, USA

[2]Department of Mechanical Engineering, Wichita State University, 1845 Fairmount, Wichita, KS 67260-0133, USA

ABSTRACT

Graphene nanoflakes in different weight percentages were added to polyurethane top coatings, and the coatings were evaluated relative to exposure to two different experimental conditions: one a QUV accelerated weathering cabinet, while the other a corrosion test carried out in a salt spray chamber. After the exposure tests, the surface morphology and chemical structure of the coatings were investigated via atomic force microscopy (AFM) and Fourier transform infrared (FTIR) imaging. Our results show that the addition of graphene does in fact improve the resistance of the coatings against ultraviolet (UV) degradation and corrosion. It is believed that this process will improve the properties of the polyurethane top coating used in many industries against environmental factors.

INTRODUCTION

Polyurethane (PU) is one of the main coatings used in the aircraft and many other industries. PU has important applications in coatings because of its outstanding properties, such as high tensile strength, chemical and weathering resistance, good processability, and mechanical properties [1–3]. However, since PU is an organic coating and subject to deterioration, its degradation has been investigated for years [1, 4–10]. The three critical factors for environmental degradation are ultraviolet (UV) light, water, and oxygen [8, 10–19]. The polymeric material generally degrades when it is exposed to these environmental influences. UV irradiation irreversibly changes the chemical structure of films and thus affects both the physical properties—loss of gloss, yellowing, blistering, cracking, and so forth—and the mechanical properties—loss of tensile strength, brittleness, changes in glass transition temperature (T_g), and so forth [1]. In the coating degradation theory [4, 20], the presence of oxygen generates hydroperoxides, which in turn accelerates photodegradation of the polyurethane. The presence of water molecules is also another factor accelerating this process. Unsaturated bonds of polymers can be activated first under UV exposure and then react with oxygen to generate carbonyl or peroxide groups. This results in the formation of carbonyl, peroxide, ketone, or aldehyde groups near the coating surface. Weathering degradation of PU can be reduced with the help of novel approaches.

An assortment of approaches can be undertaken to protect coatings against weathering conditions and radiation failures [7, 21–26]. To reduce UV degradation, UV screeners [27] are inserted into the bulk polymer. In general, these additives absorb UV light by themselves and minimize the amount of UV light absorbed on the surface of the polymer. UV screeners are high in cost and degrade because of the heavy UV exposure [7, 22,23, 28].

Applying nanomaterials to coatings, a recent practice for newly developed coating systems, has shown better performance. Nanomaterials were used to improve the properties of the coating because the nanomaterials have exceptionally high surface area-to-volume ratio which gives rise to exceptional properties in the new products. Since the aspect ratio is high, the addition of a small weight percentage of nanomaterials is sufficient to achieve the desired properties. Therefore, the properties can be drastically improved with negligible increase in weight. The nanoadditive that was used in preparing the nanocomposite coating was graphene, which was in the form of sheets having a thickness (Z-dimension) in the nanolevel (<10 nm).

Graphene is the most basic form of carbon. It is composed of sp2 bonded carbon atoms arranged in hexagonal pattern in a 2D plane. The lattice of graphene consists of two triangular shaped sublattices. The sublattices are overlapped in such a way that carbon atom from one atom sublattice is at the centroid of the other sublattice. The distance between two centroid atoms is 1.42 Å [29, 30]. Each carbon atom has one s-orbital and two p-orbitals, which result in exceptional mechanical strength of the graphene sheet. The versatile and unique properties of graphene have made it a very popular material for applications in different fields such as electronics, optics, sensors, and mechanics. Therefore, the coating system containing graphene has advantages compared to current nanoparticle-based polyurethane coatings since graphene absorbs most of the incident light and prevents degradation of the polyurethane coatings by solar irradiation. In addition, graphene provides hydrophobicity on the surfaces of the coatings and slows down the degradation process of the coatings from environmental influences, such as water, UV light, and oxygen. The insertion of nanoparticles is one of the most promising methods to improve both mechanical and weathering properties of polymeric coatings [24, 25].

Applications of nanomaterials in coatings are a recent practice for newly developed coating systems with better performance. The nanoparticles used in coatings are SiO_2 [21, 24], TiO_2 [25, 31, 32], ZnO [27], Al_2O_3 [33], and ZnS [26]. Selection of the nanoparticles is based on the inherent properties they possess. The improvement of mechanical and weathering properties is a result of the much greater surface area-to-volume ratio of the nanomaterial. Two of the nanoparticles listed above, titania and zinc oxide, are used as UV blocking agents [27]. Although they improve the antiweathering properties of the coatings, anatase titania nanomaterials produce free radicals due to their photocatalytic properties. Recently, Mirabedini et al. [31] studied the weathering performance of polyurethane nanocomposite coatings through suppressing the photocatalytic activity of titania modified with silane molecules. However, considering the mechanism of environmental weathering, water is one of the factors to accelerate the process and should be considered. In addition, to improve mechanical robustness and anticorrosion properties of the polyurethane, it is important to select the intrinsic properties of the nanomaterials. Therefore, graphene is an ideal candidate for this improvement since it absorbs most of the light and provides hydrophobicity for repelling water.

Our recent work [29] has proved that the insertion of graphene into polymer film increases water repellency. Graphene possesses exceptional mechanical strength (~0.5 TPa) [34], nanoscale dimensions, [34] and hydrophobicity, [29] and is cost effective. In addition, graphene nanomaterials absorb most of the incident light, and the insertion of small amounts of graphene nanomaterials does not affect the transparency of the film [34]. Therefore, graphene may be an ideal material to enhance mechanical and anticorrosion properties of top-layer coatings. To date, there have been no reports on enhanced anticorrosion of polymeric coatings associated with graphene. Therefore, the weathering performance of polyurethane coatings was investigated with the addition of varied amounts of graphene. Based on our hypothesis, the weathering process can be prevented and/or retarded by the use of graphene since it absorbs incident light, provides mechanical durability, and increases hydrophobicity for the top coating.

EXPERIMENTAL

Materials

Glass fiber-reinforced composite coupons 2.5 cm × 5.0 cm in size were used as test samples. We employed the prepreg lay-up technique combined with the vacuum-bagging process to fabricate the coupons. Epoxy primer was used to paint the base coat on the composites. Epoxy adduct is a polyamine-based compound that works as a hardener in conjunction with the primer. Polyester urethane-based compound shows excellent bonding with epoxy primers, so it was used as the top coat. A hardener was employed in conjunction with the top coat. Nanosize graphene platelets were purchased from Angstron Materials, product number N008-100-N. At least 80% of graphene platelets had a Z-dimension < 100 nm.

Methods

The coating system employed on a composite substrate generally consists of two layers: an epoxy-based primer and a polyurethane-based top coat. The glass fiber-reinforced plastic (GFRP) specimens used as substrates were first coated with a base primer. Next, top coats containing the nanoadditives in different weight percentages were applied. The specimens were degraded by exposing them to two different experiments conditions: one using a QUV accelerated weathering tester, and the other a corrosion test carried out in a Singleton salt spray chamber. After exposure, the surface morphology and chemical structure of the coatings were investigated with atomic force microscopy (AFM) and Fourier transform infrared (FTIR) imaging.

Graphene nanoflakes were mixed well with the polyester urethane top coat via 30 minutes of probe sonication followed by two hours of high-speed mechanical agitation. The top coats were spray-coated onto the composite samples using a Preval spray gun. The spray process was controlled to limit the thickness of the coating to 1.5 mils (~40 microns). The samples were air-dried at room temperature for 24 hours. The total thickness of the coating system (base primer + top coat) was around 3 mils (~75 microns). To ensure uniform testing conditions, the aim during the coating process was to maintain uniform coating thickness on all test specimens. We maintained the coating thickness of 3 mils (~75 microns) on all the specimens. The thickness of the coating was measured using a coating thickness gauge (PosiTest DFT Combo).

To prove our hypothesis, the samples coated with polyurethane top layers with various graphene nanoflake concentrations between 0% and 6% were prepared. Briefly, graphene nanoflakes were mixed well with the polyurethane top coat via ultrasonication. The glass fiber-reinforced plastic (GFRP) specimens were first coated with the base primer. Then, the top coats were spray-coated onto the composite samples using a Preval spray gun. The spray process was controlled to limit the thickness of the coating to 1.5 mils (~40 microns). The total thickness of the coating system (base primer + top coat) was around 3 mils (~75 microns). By simulating sunlight and rain or dew, the samples were alternatively placed in the UV chamber and salt-corrosion chamber in intervals of 12 hours. This process was followed for 20 days. The samples were characterized for change after 0, 4, 8, 12, 16, and 20 days.

The glass fiber-reinforced plastic specimens were first coated with a base primer. The standard base coat was prepared by mixing the epoxy primer with an epoxy adducts. The primer and the curing agent were mixed in a 1 : 1 weight ratio. The mixture was then slowly stirred for 15 minutes induction time before being applied onto the specimens. After the base primer had cured on the specimens, the top coat was then applied. The standard top coat was prepared by mixing a white paint with the hardener. The top coat and the hardener were mixed in a 1 : 1 weight ratio. The paint mix was given a 30-minute induction time for optimum application performance. The nanocomposite top coat was prepared by adding the graphene to the standard base coat in weight percentages of 0%, 2%, 4%, and 6%. The nanoadditives were added to the paint. Then, the mixture was stirred by a stir bar in an alternating magnetic field at room temperature for two hours. The mixture was then placed in a sonicator for 30 minutes to ensure good dispersion. The hardener was then added and stirred for five minutes using a glass stirrer.

A UV chamber in which UV light of wavelength causes degradation of the coating surface was used to simulate the environmental degradation conditions. The test was carried out according to ASTM D4587-09, which prescribes the standard procedure for UV-condensation exposures of paint and related coatings. The UV chamber has an arrangement of UVA-340 lamps, which can produce the UV spectrum. The UV chamber used in our testing was purchased from Q-Panel Company and is known as a QUV accelerated weathering tester.

In addition to UV light, other environmental factors, such as rain, humidity, and atmospheric pollutants, can also cause degradation of the coating surface. The salt spray chamber was therefore used to replicate these conditions. The corrosion test was carried out in a Singleton salt spray chamber according to ASTM B117-09. The reservoir of the salt spray chamber was filled with 5% salt solution as prescribed by standard ASTM B117-09. The temperature inside the salt fog chamber was maintained at 32 ± 3°C. The samples were placed on a rack at a 15-degree angle, pH 6.8–7.2, and average fog concentration of 1.2 mL/h. The samples were alternatively placed in the UV chamber and corrosion chamber in intervals of 12 hours. The UV chamber replicated the dry atmospheric conditions while the salt fog chamber replicated wet atmospheric conditions. This process was followed for 20 days. It is the alternating cycles of wet and dry conditions that lead to the formation of chemical compounds such as aldehydes, ketones, and peroxides inside the coatings which lead to blister formation in coatings.

Kim and Urban [34] reported the photodegradation of polyurethane coatings. The UV light causes scission of the urethane group, and the presence of the oxygen causes oxidation of the central CH_2 group between the aromatic rings resulting in the formation of carbonyl, peroxide, ketone, or aldehyde groups near the coating surface [35–39]. The soluble oxidation products dissolve in water and get absorbed into the coating during the wet period. However, during the dry period, the oxidation products are not able to escape, resulting in adsorption. The alternate cycles of absorption and adsorption produce an osmotic pressure which blisters the surface of the coating.

In order to study the degradation mechanism of the polyurethane coating, an attenuated total reflectance (ATR) FTIR study of the test samples was conducted after the UV exposure. A detailed study of the FTIR plots was carried out using a Thermal Nicolet Magna 850 IR spectrometer. In order to conduct the ATR-FTIR, a modular attachment called the Nicolet NIC-Plan was attached to the original Thermal Nicolet Magna 850 IR spectrometer.

The surface properties of the films were investigated via an optical water contact angle goniometer, which was purchased from KSV Instruments Limited, Model CAM 100. The goniometer has a built-in camera along with a software interface, which automatically snaps the profile of the liquid on the surface. The software then uses an applicable formula (Young-Dupre's equation) to automatically determine the water contact angle. AFM was used to image the surfaces of the coating before and after UV and corrosion exposure. A micro- or nanosized cantilever with a silicon or silicon nitride sharp tip (probe) was utilized to scan the specimen surface. The AFM used in our characterization process was a MFP-3D-SA Stand Alone AFM purchased from Asylum Research.

RESULTS AND DISCUSSION

In this study, 50 × 50 µm square sections on the surface of the coatings were studied under the microscope to observe their topographical changes as a result of degradation at micro- and nanoscales. The surface roughness of the polyurethane coatings with 0% and 2% graphene was around 20 nm. The polyurethane surface became rougher, and its transparency was lower when the amount of graphene reached 6%. The coatings with varied amounts of graphene showed excellent anticorrosion performance. The coatings with 0% and 2% graphene addition are discussed in detail.

Figure 1(a) shows the AFM images of the surface of a glass fiber-reinforced composite specimen with the polyurethane top coat containing 0% graphene inclusions after four days of UV and corrosion exposure. It can be clearly seen that dark pits have been formed on the surface, and the depth of pits, from section analysis of the AFM image, is around 50 nm. The diameters are between 5 µm and 20 µm. Also, the crack growth advances; the crack length at this stage is around 30 µm and the average width is around 40 nm. Figure 1(b) shows the AFM images of the surface of glass fiber-reinforced composite coated with the polyurethane containing 0% graphene after 12 days of UV and corrosion exposure. The crack formation on the polyurethane coating suggests advancement in the degradation process. The number of pits that have been formed on the surface have increased exponentially. The increase in porosity of the surface will lead to a further increase in the rate of degradation since it is easier for the moisture to seep into the coating. The larger scale pits (dark spots) are also visible on the PU coatings after the UV and salt spray degradation.

The Addition of Graphene to Polymer Coatings for Improved...   117

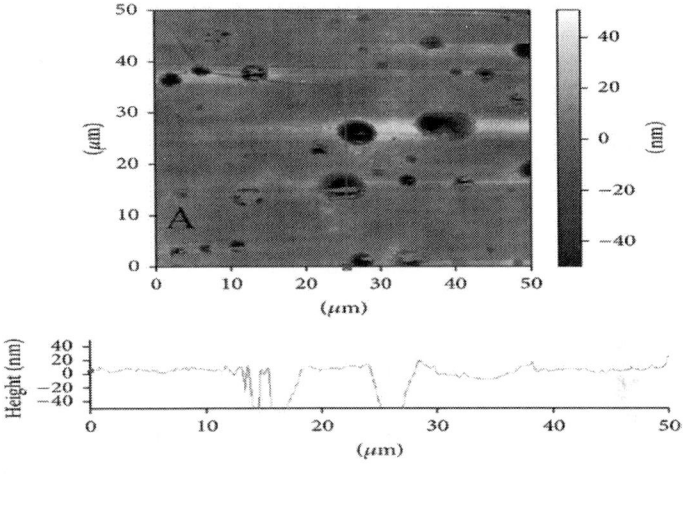

(a)

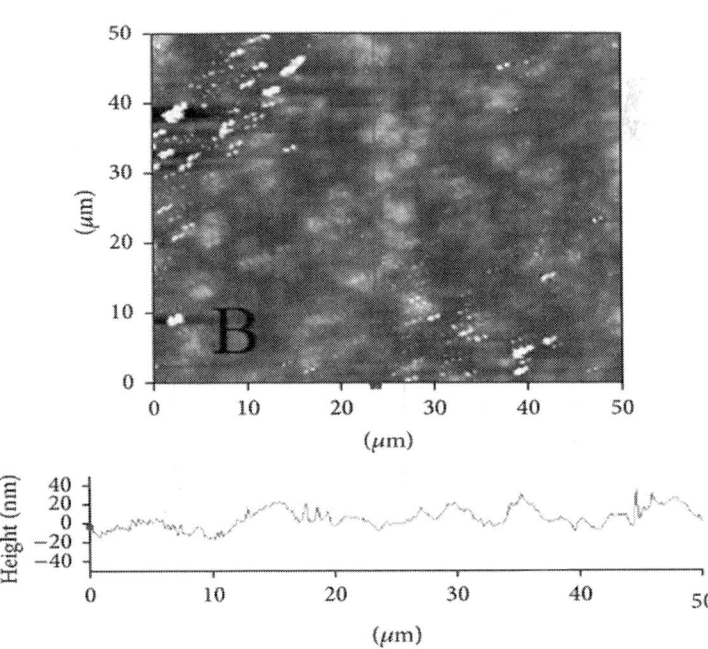

(b)

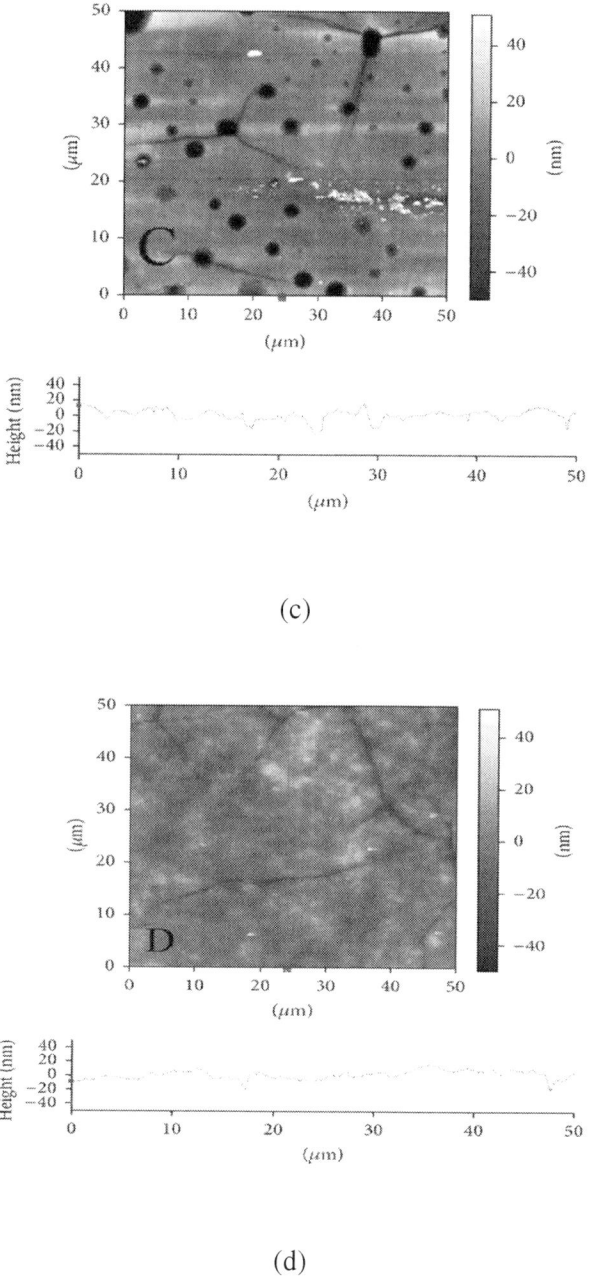

Figure 1: AFM images showing surface morphology changes of polyurethane coating with/without graphene after UV exposure and salt spray tests: (a)

AFM image of polyurethane coatings after 4 days of UV exposure and salt spray tests. (b) AFM image of polyurethane coatings with 2% of graphene after 4 days of UV exposure and salt spray tests. (c) AFM image of polyurethane coatings after 12 days of UV exposure and salt spray tests. (d) AFM image of polyurethane coatings with 2% of graphene after 12 days of UV exposure and salt spray tests.

In contrast, from the AFM image of the polyurethane (Figure 1(c)) coatings with 2% graphene under four days of UV and corrosion exposure, no visible scale pits or cracks are observed. However, Figure 1(d) shows that there are formations of cracks on the AFM images of the surface of the composite coated with polyurethane containing 2% graphene after 12 days of UV and corrosion exposure. When compared to Figure 1(b), it can be seen that the coating containing 2% graphene displayed superior properties because even after 12 days of combined UV and corrosion exposure, there is very little progress in the degradation. The only major visible sign of degradation is the presence of cracks, though without any pit formation, as seen in Figure 1(d). Although the polyurethane coatings with the addition of 4% and 6% graphene show excellent antiweathering performance, the surfaces of the coatings are rough, and the transparency of the coatings is reduced. These kinds of phenomena are also discussed in connection with other nanocomposite polyurethane coatings [31,32].

Overall, the AFM microscopy has served as a good visual indicator of the degradation process. Two conclusions can be drawn from the AFM images. First, it has been shown that a nanoadditive-like graphene, when added to a coating, can act like a reinforcement that binds the pigment cells and increases the resistance of the coating against environmental factors, such as UV degradation and corrosion. Graphene absorbs all of the light and provides hydrophobicity. Second, it has reestablished the mechanism of coating degradation, which proceeds through the formation of blisters, pits, and cracks on the surface, resulting in the loss of coating properties.

The surface properties of the polyurethane coatings were investigated for hydrophobicity. After the addition of graphene, the hydrophobicity of the coatings increased. It can be seen in Figure 2 that the 0% polyurethane coating suffers the largest decrease in contact angle with UV exposure time. At 0 days of UV exposure, the coating has a contact angle of 77.03°. However, at the end of the test cycle (20 days), the contact angle falls to 52.44°. The contact angle has decreased by almost 32%. However, the coatings with graphene show that the contact angle is higher than 80 degrees. The contact angles of polyurethane coatings decrease faster than those of coatings with added graphene. The addition of graphene in different percentages does improve the resistance against UV degradation, which is reflected by the smaller reduction in contact angle of the samples containing graphene inclusions. The coating containing 2% graphene seems to give the best results, as shown in Figure 2. The coated sample containing 2% graphene had an initial contact angle of 87.09°. After 20 days of UV exposure, the contact angle decreased by around 10 degrees. It was seen that any further increase in weight percentage of graphene does not significantly improve the resistance against UV degradation. In fact, an increase in weight percentage of graphene beyond 6% might deteriorate the properties of the coatings. In comparison with polyurethane coatings with graphene addition, the polyurethane coating degraded quickly, and pits and cracks formed on the surface of the coatings. The holes and cracks, as can be seen in the AFM images, are the main reasons for decreasing the contact angle. This can be related to capillary action as well.

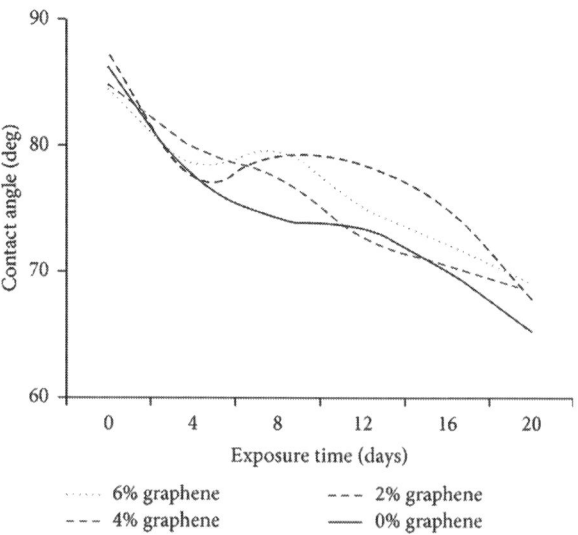

Figure 2: Change in contact angle values of UV exposed PU top coatings containing different percentages of graphene inclusions.

To understand the mechanism of the UV and corrosion damage of the coatings, ATR-FTIR was applied to further investigate any changes at the molecular level. The FTIR spectrum (Figure 3) showed some changes in the properties of polyurethane coating as a function of UV exposure. The aim of carrying out the FTIR studies was to test the hypothesis that the degradation of the PU coatings may occur due to UV exposure by the formation of certain types of compounds, such as carbonyl, aldehydes, ketones, and peroxide groups, by the scission of chains in polyurethane.

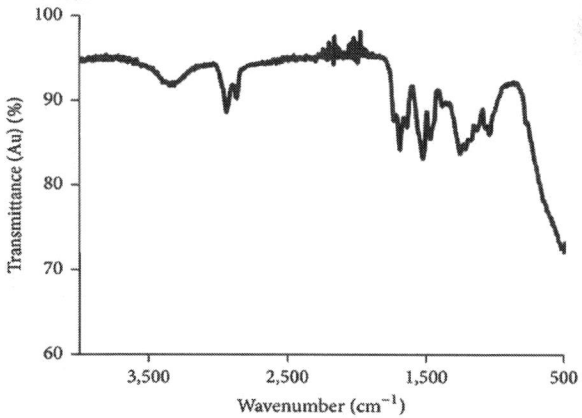

Figure 3: ATR-FTIR spectrum of PU containing 0% graphene after 20 days of UV and corrosion exposure.

Figure 3 shows the ATR-FTIR spectrum of the PU top coat containing 0% graphene. The spectrum shows both positive peaks and negative peaks. The positive peaks may indicate the formation of new structures, while negative peaks may show the loss of breakage of structures as a result of the UV irradiation. At the left side of the graph (Figure 3), the ATR-FTIR spectrum reveals that a band is formed at 3304.48 cm^{-1}. This peak can be a characteristic of stretching of the N–H group, which may suggest the formation of polyurea.

Towards right on the graph (Figure 3), a peak is seen at 2928.04 cm^{-1}, which may indicate the asymmetric and symmetric stretching of the CH_2 group. Moving into the region between 2000 and 1700 cm^{-1}, a vibration signal can be observed. This may be due to the vibration of the C=O bond. The peak at 1723.92 cm^{-1} indicates the stretching of the

C=O bond. Another strong indicator of the presence of polyurea can be the peak at 1681.83 cm^{-1}. A peak at 1483.48 cm^{-1} may indicate a decrease in the C–H group, and a peak at 1248.81 cm^{-1} may indicate a reduction in C–O group. Finally, a peak seen at 1037.10 cm^{-1} may confirm the formation of an ester group. The decrease in C–H group and C–O directly points to chain scission of polyurethane.

Figure 4 shows the ATR-FTIR spectrum of the PU top coat containing 6% graphene. This graph is mostly identical to the previous one, indicating a similar mechanism of UV degradation. Once again a peak is seen at 3323.44 cm^{-1}, possibly indicating the stretching of NH group. The twin peaks at 2928.05 cm^{-1} and 2858.28 cm^{-1} may indicate the stretching of CH$_2$ group. The vibration signal between 2200 cm^{-1} and 1700 cm^{-1} is followed by a series of peaks between 1700 cm^{-1} and 1400 cm^{-1}. The peaks may illustrate that chain scission possibly takes place as a result of the UV exposure. A single peak at 1248.52 cm^{-1} may be the indication of the reduction of C–O group. However, once again the absence of any negative peaks beyond 1100 cm^{-1} may hint towards smaller ester concentration. The decreased ester concentration strengthens the argument that graphene may possibly improve the resistance against the UV degradation by absorbing a substantial amount of the incident radiation. More studies will be needed to confirm this statement.

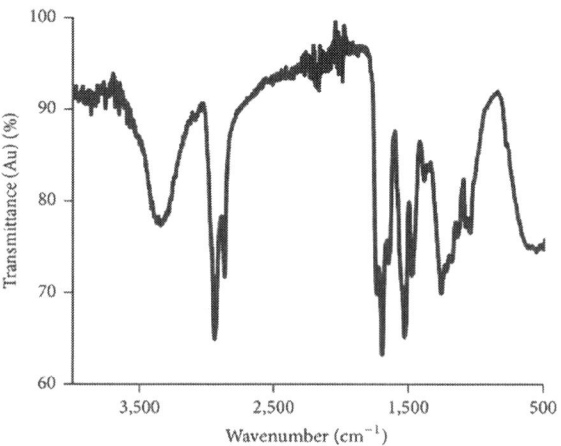

Figure 4: ATR-FTIR spectrum of PU containing 6% graphene after 20 days of UV and corrosion exposure.

CONCLUSIONS

Graphene in different weight percentages was added to polyurethane coatings, and subsequently tests were conducted to check the variation in properties of the coatings. Test samples were prepared by the addition of 0%, 2%, 4%, and 6% weight percentages of graphene into standard polyurethane coatings. FTIR spectroscopy, AFM examination, and water contact angle tests were performed to quantify the variation in properties. The tests confirmed the hypothesis that addition of graphene does in fact improve the resistance against UV degradation and corrosion. The polyurethane coating containing 2% graphene showed greatly improved performance as compared to the standard polyurethane coating, since graphene provides hydrophobicity, absorbs incident light, and improves mechanical robustness of the coatings. Also, the detailed FTIR analysis reinforced the hypothesis that degradation of polyurethane coatings occurred due to the formation of certain water soluble compounds, such as carbonyls, aldehydes, ketones, and peroxides. Through a time series study of the AFM images at different stages of the UV and corrosion tests, the progression of degradation was explained in detail by the formation and enlargement of blisters, pits, and cracks. Overall, the research has provided a detailed overview of the mechanism of coating degradation and also suggested a means to arrest or decrease the rate of this degradation by the use of a nanoadditive, namely, graphene.

ACKNOWLEDGMENTS

The authors would like to acknowledge the Department of Energy for the financial support (DE-EE0004167) and the Department of Chemistry at Wichita State University for the technical support of this work.

REFERENCES

1. H. Aglan, M. Calhoun, and L. Allie, "Effect of UV and hygrothermal aging on the mechanical performance of polyurethane elastomers," Journal of Applied Polymer Science, vol. 108, no. 1, pp. 558–564, 2008.

2. G. Oertel and L. Abele, Polyurethane Handbook: Chemistry, Raw Materials, Processing, Application, Properties, Hanser, St. Louis, Mo, USA, 2nd edition, 1994.

3. S. Ramakrishna, J. Mayer, E. Wintermantel, and K. W. Leong, "Biomedical applications of polymer-composite materials: a review," Composites Science and Technology, vol. 61, no. 9, pp. 1189–1224, 2001.

4. X. F. Yang, D. E. Tallman, G. P. Bierwagen, S. G. Croll, and S. Rohlik, "Blistering and degradation of polyurethane coatings under different accelerated weathering tests," Polymer Degradation and Stability, vol. 77, no. 1, pp. 103–109, 2002.

5. X. F. Yang, C. Vang, D. E. Tallman, G. P. Bierwagen, S. G. Croll, and S. Rohlik, "Weathering degradation of a polyurethane coating," Polymer Degradation and Stability, vol. 74, no. 2, pp. 341–351, 2001.

6. A. Ludwick, H. Aglan, M. O. Abdalla, and M. Calhoun, "Degradation behavior of an ultraviolet and hygrothermally aged polyurethane elastomer: Fourier transform infrared and differential scanning calorimetry studies," Journal of Applied Polymer Science, vol. 110, no. 2, pp. 712–718, 2008.

7. C. Merlatti, F. X. Perrin, E. Aragon, and A. Margaillan, "Natural and artificial weathering characteristics of stabilized acrylic-urethane paints," Polymer Degradation and Stability, vol. 93, no. 5, pp. 896–903, 2008.

8. S. P. Pappas, "Weathering of coatings—formulation and evaluation," Progress in Organic Coatings, vol. 17, no. 2, pp. 107–114, 1989.

9. B. M. D. Fernando, X. Shi, and S. G. Croll, "Molecular relaxation phenomena during accelerated weathering of a polyurethane coating," Journal of Coatings Technology Research, vol. 5, no. 1, pp. 1–9, 2008.

10. D. R. Bauer, "Degradation of organic coatings. I. Hydrolysis of melamine formaldehyde/acrylic copolymer films," Journal of Applied Polymer Science, vol. 27, no. 10, pp. 3651–3662, 1982.

11. W. Schnabel, Polymer Degradation, Principles and Practical Applications, Macmillan, New York, NY, USA, 1981.

12. R. R. Blakey, "Evaluation of paint durability—natural and accelerated," Progress in Organic Coatings, vol. 13, no. 5, pp. 279–296, 1985.
13. G. Z. Xiao and M. E. R. Shanahan, "Water absorption and desorption in an epoxy resin with degradation," Journal of Polymer Science, Part B, vol. 35, no. 16, pp. 2659–2670, 1997.
14. A. M. Morrow, N. S. Allen, and M. Edge, "Photodegradation of water-based acrylic coatings containing silica," Journal of Coatings Technology, vol. 70, no. 880, pp. 65–72, 1998.
15. D. R. Bauer, "Network formation and degradation in urethane and melamine-formaldehyde crosslinked coatings," Polymeric Materials, vol. 56, pp. 91–95, 1987.
16. D. R. Bauer, "Melamine/formaldehyde crosslinkers: characterization, network formation and crosslink degradation," Progress in Organic Coatings, vol. 14, no. 3, pp. 193–218, 1986.
17. D. R. Bauer, R. A. Dickie, and J. L. Koenig, "Cure and photodegradation of two-package acrylic/urethane coatings," Industrial and Engineering Chemistry Product Research and Development, vol. 25, no. 2, pp. 289–296, 1986.
18. J. L. Gerlock, H. Van Oene, and D. R. Bauer, "Nitroxide kinetics during photodegradation of acrylic/melamine coatings," European Polymer Journal, vol. 19, no. 1, pp. 11–18, 1983.
19. D. R. Lefebvre, K. M. Takahashi, A. J. Muller, and V. R. Raju, "Degradation of epoxy coatings in humid environments: the critical relative humidity for adhesion loss," Journal of Adhesion Science and Technology, vol. 5, no. 3, pp. 201–227, 1991.
20. B. G. Rånby and J. F. Rabek, Photodegradation, Photo-Oxidation, and Photostabilization of Polymers: principles and applications, John Wiley & Sons, New york, NY, USA, 1975.
21. S. Zhou, L. Wu, W. Shen, and G. Gu, "Study on the morphology and tribological properties of acrylic based polyurethane/fumed silica composite coatings," Journal of Materials Science, vol. 39, no. 5, pp. 1593–1600, 2004.
22. C. Decker, F. Masson, and R. Schwalm, "Weathering resistance of waterbased UV-cured polyurethane-acrylate coatings," Polymer Degradation and Stability, vol. 83, no. 2, pp. 309–320, 2004.

23. B. H. Lee and H. J. Kim, "Influence of isocyanate type of acrylated urethane oligomer and of additives on weathering of UV-cured films," Polymer Degradation and Stability, vol. 91, no. 5, pp. 1025–1035, 2006.
24. M. Sangermano, G. Malucelli, E. Amerio, A. Priola, E. Billi, and G. Rizza, "Photopolymerization of epoxy coatings containing silica nanoparticles," Progress in Organic Coatings, vol. 54, no. 2, pp. 134–138, 2005.
25. S. M. Mirabedini, M. Mohseni, S. PazokiFard, and M. Esfandeh, "Effect of TiO_2 on the mechanical and adhesion properties of RTV silicone elastomer coatings," Colloids and Surfaces A, vol. 317, no. 1-3, pp. 80–86, 2008.
26. C. Lü, Z. Cui, Z. Li, B. Yang, and J. Shen, "High refractive index thin films of ZnS/polythiourethane nanocomposites," Journal of Materials Chemistry, vol. 13, no. 3, pp. 526–530, 2003.
27. A. Ammala, A. J. Hill, P. Meakin, S. J. Pas, and T. W. Turney, "Degradation studies of polyolefins incorporating transparent nanoparticulate zinc oxide UV stabilizers," Journal of Nanoparticle Research, vol. 4, no. 1-2, pp. 167–174, 2002.
28. M. M. Jalili and S. Moradian, "Deterministic performance parameters for an automotive polyurethane clearcoat loaded with hydrophilic or hydrophobic nano-silica," Progress in Organic Coatings, vol. 66, no. 4, pp. 359–366, 2009.
29. C. Soldano, A. Mahmood, and E. Dujardin, "Production, properties and potential of graphene," Carbon, vol. 48, no. 8, pp. 2127–2150, 2010.
30. N. S. Allen, M. Edge, A. Ortega et al., "Degradation and stabilisation of polymers and coatings: nano versus pigmentary titania particles," Polymer Degradation and Stability, vol. 85, no. 3, pp. 927–946, 2004.
31. S. M. Mirabedini, M. Sabzi, J. Zohuriaan-Mehr, M. Atai, and M. Behzadnasab, "Weathering performance of the polyurethane nanocomposite coatings containing silane treated TiO_2 nanoparticles," Applied Surface Science, vol. 257, no. 9, pp. 4196–4203, 2011.
32. S. K. Dhoke, T. J. M. Sinha, and A. S. Khanna, "Effect of nano-Al_2O_3 particles on the corrosion behavior of alkyd based waterborne

coatings," Journal of Coatings Technology Research, vol. 6, no. 3, pp. 353–368, 2009.
33. R. Asmatulu, M. Ceylan, and N. Nuraje, "Study of superhydrophobic electrospun nanocomposite fibers for energy systems," Langmuir, vol. 27, no. 2, pp. 504–507, 2011.
34. H. Kim and M. W. Urban, "Molecular level chain scission mechanisms of epoxy and urethane polymeric films exposed to UV/H_2O. Multidimensional spectroscopic studies," Langmuir, vol. 16, no. 12, pp. 5382–5390, 2000.
35. R. L. Feller, Accelerated Aging: Photochemical and Thermal Aspects, The J. Paul Getty Trust, Los Angeles, Calif, USA, 1994.
36. Z. W. Wicks Jr., F. N. Jones, and S. P. Pappas, Organic Coatings: Science and Technology, John Wiley & Sons., 2nd edition, 1994.
37. C. G. Overberger and G. Menges, "Fabrication to die design," in Encyclopedia of Polymer Science and Engineering in Composites, H. F. Mark, Ed., vol. 4, John Wiley & Sons, New York, NY, USA, 1994.
38. J. Lemaire, R. Arnaud, and J. L. Gardette, "The role of hydroperoxides in photooxidation of polyolefins, polyamides and polyurethane elastomers," Pure and Applied Chemistry, vol. 55, no. 10, pp. 1603–1614, 1983.
39. R. Asmatulu, G. A. Mahmud, C. Hille, and H. E. Misak, "Effects of UV degradation on surface hydrophobicity, crack, and thickness of MWCNT-based nanocomposite coatings," Progress in Organic Coatings, vol. 72, no. 3, pp. 553–561, 2011.

Chapter 5

Recent Progress in Processing of Tungsten Heavy Alloys

Y. Şahin

Department of Manufacturing Engineering, Faculty of Technology, Gazi University, Beşevler, 06500 Ankara, Turkey

ABSTRACT

Tungsten heavy alloys (WHAs) belong to a group of two-phase composites, based on W-Ni-Cu and W-Ni-Fe alloys. Due to their combinations of high density, strength, and ductility, WHAs are used as radiation shields, vibration dampers, kinetic energy penetrators and heavy-duty electrical contacts. This paper presents recent progresses in processing, microstructure, and mechanical properties of WHAs. Various processing techniques for the fabrication of WHAs such as conventional powder metallurgy (PM), advent of powder injection molding (PIM), high-energy ball milling (MA), microwave sintering

(MW), and spark-plasma sintering (SPS) are reviewed for alloys. This review reveals that key factors affecting the performance of WHAs are the microstructural factors such as tungsten and matrix composition, chemistry, shape, size and distributions of tungsten particles in matrix, and interface-bonding strength between the tungsten particle and matrix in addition to processing factors. SPS approach has a better performance than those of others, followed by extrusion process. Moreover, deformation behaviors of WHA penetrator and depleted uranium (DU) Ti alloy impacting at normal incidence both rigid and thick mild steel target are studied and modelled as elastic thermoviscoplastic. Height of the mushroomed region is smaller $\alpha = 0.3$ for and it forms sooner in each penetrator as compared to that for. $\alpha = 0.2$

INTRODUCTION

The name of "tungsten" is derived from the Swedish term meaning "heavy stone." Tungsten has been assigned the chemical symbol W after its German name wolfram. Tungsten, the metal with the highest meting point (3422°C), has many advantages, such as high temperature strength, high creep resistance and high thermal conductivity, high electric resistance, the lowest vapor pressure, and the lowest coefficient of thermal expansion. These properties make tungsten a premium candidate for high temperature applications like, for example, in fusion reactor [1]. Another important industrial property of tungsten is its high density of 19.3 g/cm^3, which makes it an ideal material for shielding or collimating energetic - and -radiation. The disadvantage of tungsten, however, is its inherent brittleness because tungsten has a transition from brittle to ductile fracture. Its treatment is realized at temperatures that are higher than brittleness limit. This temperature varies for commercially pure tungsten (99.95%) in the interval between 300 and 400°C, in case of recrystallized tungsten around 500°C. Undesirable mixtures such as oxygen, nitrogen, and carbon significantly influence mechanical and physical properties of pure tungsten. They mainly precipitate at the grain boundaries in the form of oxides, nitrides, and carbides [2]. It is possible to reduce the transition temperature of pure tungsten by rotary sawing at temperatures 1550–1450°C, transition temperature drops at 150°C. Tungsten's strength and plastic characteristics increase with making the tungsten alloy, which can be divided into three basic groups: (a) solid solution alloys, (b) heterogeneous alloys like dispersion-hardening alloys, and (c) precipitation-hardening alloys. Alloying elements like Nb, Ta, Mo, Zr, Rh, B, C, and others influence tungsten's strength properties and its plastic deformability. Pure tungsten has a body-centered cubic (BCC) material whose single crystals are virtually elastically isotropic at low pressures. The major uses for pure tungsten are in wire form in electrical lamps and electronic vacuum tubes, glass-metal sealing rods, cathode materials, electrical contacts, and heating elements [3]. Tungsten powder with an average particle size from 0.5 to 40 μm is used for producing tungsten carbide as cutting tools, milling tools, and mining drills. Furthermore, since the late 1960s, tungsten carbide has been considered as an anode material for hydrogen or methanol fuel cells [4, 5].

Tungsten heavy alloys (WHAs) are a typical class of two-phase composites consisting of nearly spherical tungsten particles embedded in a ductile matrix phase with lower melting point elements such as nickel, iron, copper, and cobalt [6]. The spheroidized phase microstructure is essentially pure tungsten, which is surrounded by a metallic binder phase containing some dissolved tungsten particles over 80 wt.% [7]. WHAs have been processed through PM route for some specific applications like mechanical alloying and infiltration since 1980s [8]. These heavy alloys contain mainly pure tungsten as principal phase in association with a matrix phase containing transition metals. Ni to Cu ratio in the heavy alloys can range from 3.2 to 4.1. Price et al. [9] are the first to propose the Ni-Cu as the binder for tungsten heavy alloys. Over the last several years, these alloys have been extensively investigated for densification mechanism and microstructural evaluation and properties [10, 11]. WHAs are usually fabricated by liquid-phase sintering (LPS) process from micron-sized W-Ni-Fe or W-Ni-Cu elemental powder mixtures at a temperature above 1460°C [12]. During LPS, the nickel alloy melt, which has a high solubility for tungsten, dissolves the tungsten-tungsten bonds that are formed during solid-state sintering. When the solid bonds are penetrated, particle rearrangement and pour filling occur, resulting in rapid densification of the compact. Therefore, tungsten solubility and tungsten grain shapes retention are two main concerns in determining the mechanical properties of sintered heavy alloys. Use of prealloyed powder improves homogeneity. The homogenization process also accelerates sintering and promotes densification. Ramakrishnan and Upadhayaya [13] and Kuzmic [14] study the effects of composition, temperature, and rapid cooling on densification and microstructure of the heavy alloys. The heavy alloys can be produced in a variety of near-net shapes, but the most frequently encountered shapes are those of cylindrical rods. After sintering, the final microstructure consists of coarsened-large spherical BCC tungsten grains dispersed in FCC solidified matrix. The final tungsten grain size is typically about 20–60 μm in diameter. The properties of LPS materials in general are degraded by prolonged final stage of sintering time. Hence, short sintering times are preferred in practice. However, the liquid-phase sintered WHAs are susceptible to distortion and slumming due to the large difference between tungsten and liquid matrix. Thus, understanding and controlling the microstructure evaluation in terms of composition, densification, processing like heating rate, solubility properties, and the structural rigidity are of great importance to materials engineers and scientists in the processing of WHAs.

Due to the outstanding combination of the properties related to body-centered cubic tungsten phase and face-centered cubic matrix, such as high physical and mechanical properties, such as high density (16–18 g/cm^3), high strength (1000–1700 MPa), and high ductility (10–30%) [15], thermal conductivity, and good corrosion resistance, WHAs are widely used in radiation shields, counterbalance, vibration dampers, kinetic energy (KE) penetrators and rocket nozzles in space crafts [16–20]. The major disadvantage of WHAs is that they have a lesser penetration capability than depleted uranium alloys (DUs), which are another penetrator material [21]. The DUs show superior penetration performance because of self-sharpening behavior. WHAs, however, develop mushroom-like heads (blunt behavior) during penetration due to adiabatic shear deformation, which result in a lower penetration depth when compared to depleted uranium alloys. For penetrators, there are two key factors affecting the penetration ability which are good penetration and high density in addition to low heat capacity and low strain rate hardening [8, 21]. So far, most investigators focused mainly on enhancement of the mechanical properties of WHAs in order to improve the penetration capabilities for kinetic energy penetration applications. For achievement of this purpose, several techniques are proposed including grain size control of W, alloying elements like Mo, Re, solid state interring, mechanical alloying, cold working followed by recrystalline, cyclic heat treatment, and oxide dispersion strengthening and many useful experimental results have been obtained [17–24]. This paper reviews recent progress on the processing, microstructure, and mechanical properties of WHAs. In addition, the Johnson-Cook relation [25] and Zhou et al. [26] thermal softening function are reviewed to model the axisymmetric, elastic thermoviscoplastic deformations of DU and WHA rods penetrating a mild steel target.

PROCESSING TECHNIQUES

As tungsten based materials are refractory materials, melting and casting of these materials are extremely difficult. Therefore alternative-processing techniques are required, which are discussed below.

Powder Metallurgy

Powder metallurgy (P/M) is a process, in which a material powder is compacted as a green body and sintered to a net shape at elevated temperature. The basic steps consist of blending or mixing of elemental/alloy WHAs powders (a), followed by compacting the mixture in a suitable die or mold (b), and heating the resulting green compacts in a controlled atmosphere furnace so as to bond the particles metallurgically (c); this heating process is called "sintering" [27]. The resulting parts are solid bodies of material with sufficient strength and density for use in diverse fields of applications. In some times, steeps (a) and (b) are combined, which is called "hot isostatic pressing," because the powder mix is subjected simultaneously to pressure and elevated temperature. The sintered parts may be subjected to one or more secondary finishing operations like grinding and plating. The first production of WHAs and tungsten carbide materials by P/M method started in the early part of the last century. To obtain homogeneous distributions of rounded W solid phases in the alloy is important to tackle in the processing step. Solid solution alloy is produced when, for example, the element Re is added to a 93W-4.9Ni-2.1Fe alloy, inhibiting tungsten grains from precipitation growth [28]. However, adding the element Cr causes the interphases with elements W, Ni, Fe, and O to accumulate along interface alloy, resulting in lower mechanical properties of the alloys. Addition of Mo reduces the concentration of W in the liquid matrix phase during sintering and, consequently, fines the microstructure of alloys [29, 30]. Former studies indicate the liability of forming a precipitated phase with the addition of high concentration of Mo, which lead to brittleness of the alloy [31]. The coprecipitation of W, Mo, Ni, and Fe results in the formation of an intermetallic phase at the interface like W_4Mo and Ni_7Fe, which is in good agreement with the early work [32, 33]. As Mn/Ni ratio increases, W solubility in the matrix decreases due to the presence of Mn in matrix and causes refinement of W grains [34].

LPS at various temperatures [35–38] prepare microstructure and mechanical properties of WHAs with different compositions. All results are dependent on grain size, alloying content, compositions, and sintering time. The first two show increase in tensile properties and hardness in the presence of a finer W grain size. The tensile strength and elongation remarkably deteriorate at higher temperatures. The third indicates that both microstructure and mechanical properties are sensitively depending on alloy compositions, while the effect of varying W content significantly affects the microstructure of WHAs [38]. By increasing the Ni/Fe ratio, there is higher solubility for W in the matrix, giving a slightly larger grain size with a lower contiguity. The tensile strength and elongation are the highest for sintering times from 30 to 90 min, reflecting a minimum in the residual porosity [39, 40]. However, decreasing the yield strength with increasing sintering time is in agreement with the Hall-Petch behavior of 93W-4.9Ni-2.1Fe heavy alloy [41, 42]. Densification, microstructure, and mechanical properties of 90W-4Ni-6Mn heavy alloys indicate that temperature most critically influences the microstructure because sintering density decreases rapidly with increasing temperature while temperature decreases with decreasing the grain size [43]. The grain coalescence and distortion are induced by prolonging the sintering time or elevating sintering temperature [44]. In comparison to 93W-4.9Ni-2.1Fe alloy, 93W-7(Ni-Fe-Mo) heavy alloy has a higher tensile strength and a better ductility [45]. Heating rate during sintering is another parameter that affects the densification and distortion. The grain growth rate constant varies inversely with the volume fraction of liquid to 2/3 power [46]. The grain size, volume fraction of binding phase, and microhardness vary gradually due to the graded distribution of Mo [29]. The general trend is an increase in solid volume fraction, contiguity, and grain size with increasing W content [47, 48]. The strength of the alloy increases significantly due to greater hardening of the matrix phase. The strength properties result in an embrittlement behavior with decreasing the temperature [49].

Deformation properties of WHAs are studied under different conditions; the strength of the alloy increases significantly due to a greater hardening of the matrix phase and W grains are elongated significantly after deformation mechanism [50]. However, strength and elongation of WHAs are deteriorated very rapidly when tested over 650°C. The strength properties result in an embrittlement behavior with decreasing the temperature. The dynamic compression failure study indicates that shear band formation (ASB) is a failure mechanism for WHAs [51–53]. The microstructure evaluation of ASB formation is mainly based on the dynamic deformation process and the initial temperature [54–56]. ASB and penetration performance will be discussed within a more detailed way at the end of the mechanical behavior section. Powder metallurgy techniques can generally be classified into powder injection molding, mechanical alloying, microwave, and spark plasma sintering; these are discussed briefly in the following.

Powder Injection Molding

The powder injection molding (PIM) process is a combination of powder metallurgy and plastic injection molding technologies. This process offers three main advantages such as precise and reproducible components, complex shape, and high densification due to use of very fine ceramic or metal powders in the feedstock [57]. The process consists of mixing a small amount of organic polymer materials like wax polymer, polyacetal, and so forth with the desired inorganic powder, followed by granulation or pelletisation of the mixture [58]. The processing steps consist of feedstock preparation, injection molding, debinding, and consolidation [59]. The control of binder removal and maintaining its shape integrity are a challenge for the process. Unless the binder is removed gradually and the removal of the organic binder phase is completed, numerous defects can be formed when producing WHAs. The tensile properties are comparable to the conventional press and sinter alloys of a similar composition [60]. The solvent debinding temperature and thermal debinding atmosphere are found to be the principal factors affecting the mechanical properties of the alloy [61, 62]. Effect of agglomerates with different contents on the green and sintered properties of 97W-2.1Ni-0.9Fe heavy alloys for tensile specimens indicates that particle characteristics have no effect on the microstructure and mechanical properties because agglomeration remains as porous structure in the molded green compacts [63, 64]. The feedstock formed by mechanically alloyed (MAed) for W-Ni-Fe nanocrystalline composite powder and wax multicomponent binder has good flow ability and mold ability. 90W-7Ni-3Fe composites by PIM are widely investigated [65–69] mainly due to unique combination of high density, high strength, good ductility, and corrosion resistance of these materials. Solid-state sintering in the range between 1350 and 1450°C [70] produces fully densified parts with high mechanical properties and negligible distortion. The results indicate that the ultimate tensile strength exhibits the highest value at a sintering temperature of 1500°C and an isothermal holding time of 40 min [71, 72]. Micropowder injection molding (μPIM) is another new and fast-developing micromanufacturing technique for the production of metal and ceramic component.

High Energy Ball Milling

Mechanical alloying (MA), which represents a high-energy ball milling process, has established itself as a viable solid state-processing route for the synthesis of a variety of equilibrium and nonequilibrium phase mixtures. It was originally developed to produce oxide-dispersion strengthened (ODS) nickel- and iron-base super alloys for applications in aerospace industry. Benjamin [73] developed this technique around 1966. The powders are mixed in the mill along with the steel balls. This mix is then milled for the desired length of time until a steady state is reached. During this high-energy milling, the powder particles are repeatedly flattened, cold-welded, fractured, and rewelded [74]. Figure 1 shows the schematic ball movement in typical ball mills used for MA [75]. Among the laboratory mills, the SPEX shaker mills, which mill about 10 ± 20 g of powder at a time, are most commonly used for alloy screening purposes. The energy transfer to the powder particles in these mills takes place by a shearing action or impact of the high velocity balls with the powder. This transfer is governed by many parameters such as type of mill, milling speed, type, size and size distribution of balls, ball/powder weight ratio, extent of filling of the vial, dry or wet milling, temperature of milling, atmosphere in the mill, and, finally, the duration of milling [74,75]. The mechanical alloying process of 93W-5.6Ni-1.4Fe tungsten heavy alloys shows that steady state stage is reached after milling for 48 hours and the grain size is refined to 16 μm at the steady state stage. When solid-state sintered at 1300°C for 1 h in a hydrogen atmosphere, MAed WHA indicates ultrafine tungsten particles of about 3 μm in diameter with high density (above 99%) [76]. These solid-state sintered WHAs with MA powders exhibit high yield strength of about 1100 MPa due to a fine microstructure but show reduced elongation and impact energy mainly caused by a large area fraction of brittle W-W interface [77–79].

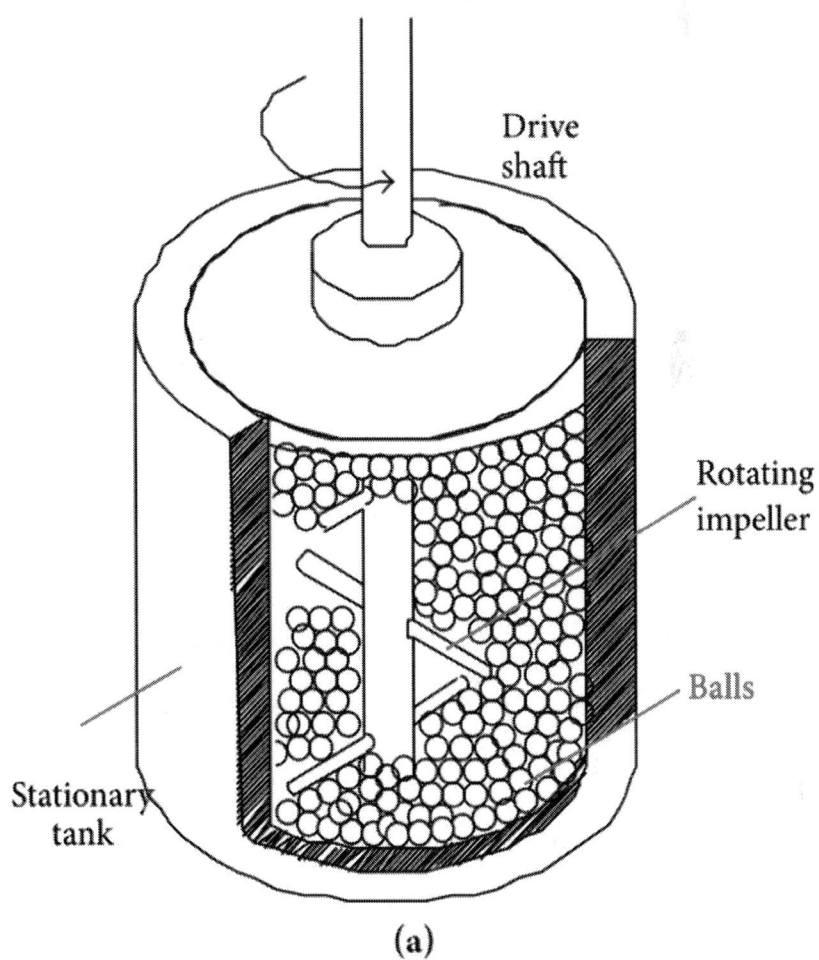

(a)

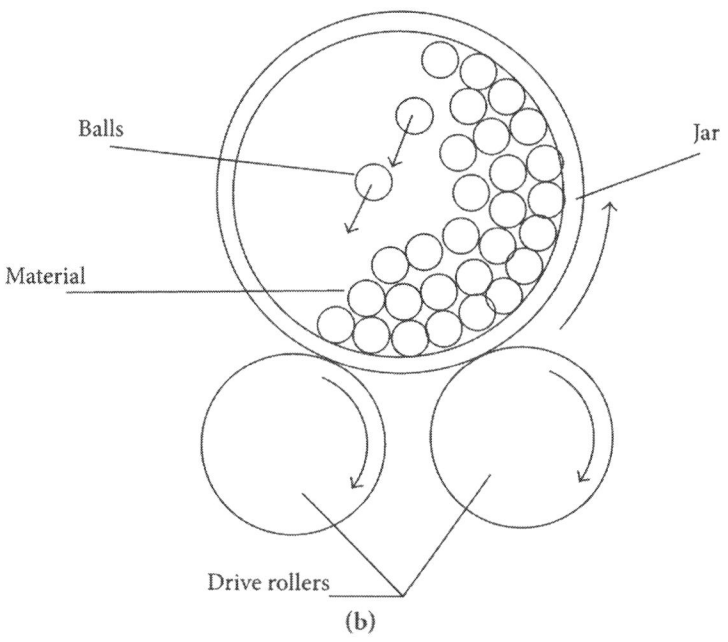

(b)

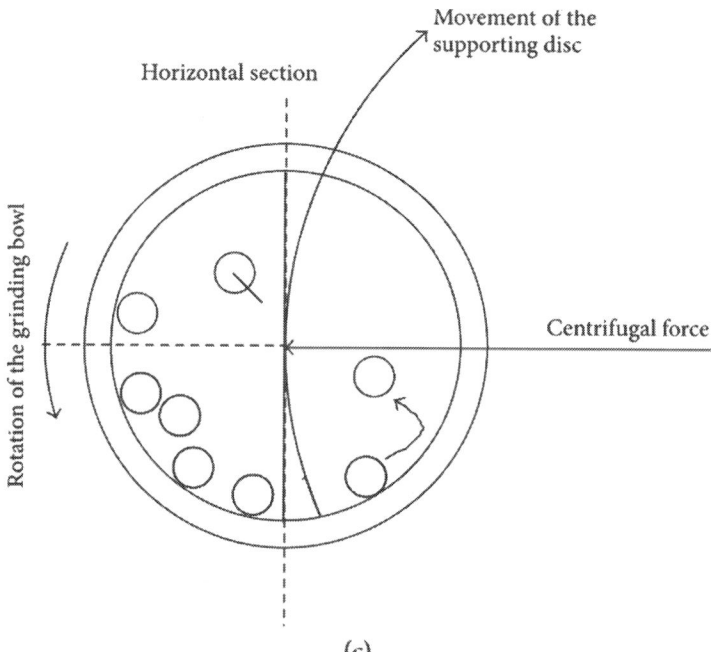

(c)

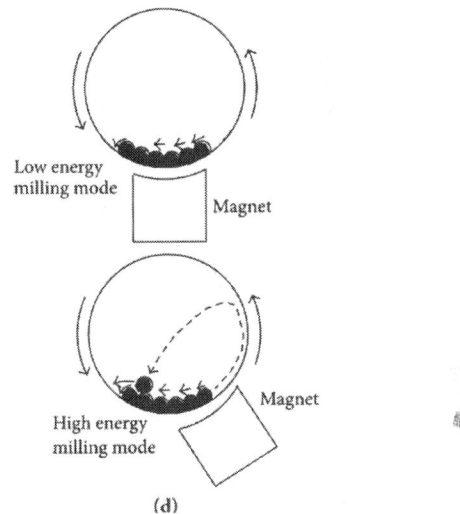

Figure 1: Schematic of ball movement in typical ball mills used for MA [75]. (a) Attritor, (b) tumbler mill, (c) planetary mill, and (d) uniball mill.

Figure 2 shows the SEM micrographs of two-stage sintered 93W-5.6Ni-1.4Fe WHA secondarily sintered at 1470°C for 4 min and 90 min after solid-state sintering at 1300°C for 1 h [77]. The composition of 93W-4.9Ni-2.1Fe alloy is mechanically alloyed (MAed) and nanocrystalline supersaturated solid solutions with grain size of 11 nm and amorphous phase are achieved during MA [80–83]. When sintered at 1150°C for 30 min, the MAed powders show homogeneous microstructure and ultrafine tungsten particles of approximately 2 μm with high density above 95%. When 90W-7Ni-3Fe nanocrystalline composite powder dispersed with Y_2O_3 addition fabricated by MA method, the maximum tensile strength obtained is 1050 MPa and the elongation is 30% [84]. Milling increases solubility of W in phase and the solubility decreases with an increase of temperature [85]. 93W-3Ni-2Fe-2Co and 93W-3.5Ni-1.5Fe-2Copre-alloyed powders with crystal cell size about 16 nm are synthesized by MA, and the hardness of these sintered WHA exhibits a trend with increasing sintering temperature and sintering time [86]. However, the nanosized powders by MA process exhibit that the average W grain size in the range of 1.7–3.0 μm is obtained. The tensile strength more than 1200 MPa is achieved at a sintering temperature of 1350°C.

(a)

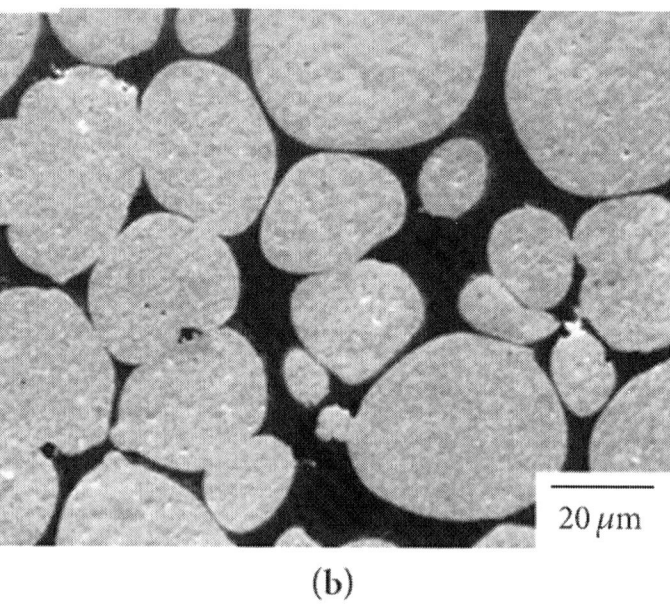

(b)

Figure 2: SEM micrograph of two-stage sintered 93W-5.6Ni-1.4Fe WHAs secondarily sintered at 1470°C for (a) 4 min and (b) 90 min after solid-state sintering at 1300°C for 1 h [78].

Significant improvements in mechanical properties of WHAs are achieved by hot extrusion, hot isostatic extrusion, and cold deformation, forged and swaging of compacts prepared from ball milled powder or powder mixtures [87–90], but the last one indicates that the mechanical properties depend on the amount of deformation imparted during swaging. Rapid heat experiments of the order of seconds are performed on a 90W-7Ni-3Fe alloy, which is cold worked by rotary swaging to 50 and 82% reduction in diameter [90]. Refined and elongated W grains into small round ones are obtained and increased deformation enhances the process of partition of W grains. Figure 3 shows a SEM micrograph of the longitudinal microstructure of the liquid-phase sintered and hot-hydrostatically extruded 93W-4.9Ni-2.1Fe alloy samples [91, 92]. W particles in the alloy are elongated along the extrusion direction. Figure 3(b) shows that the higher the average aspect ratio of the elongated W particles the bigger the extrusion ratio (2.86 : 1) used for hot-hydrostatic extrusion (Figure 3(b)). The ultimate tensile strength and hardness of the alloy were enhanced significantly and the increment in tensile strength and hardness is in proportion to the extrusion ratio [92].

(a)

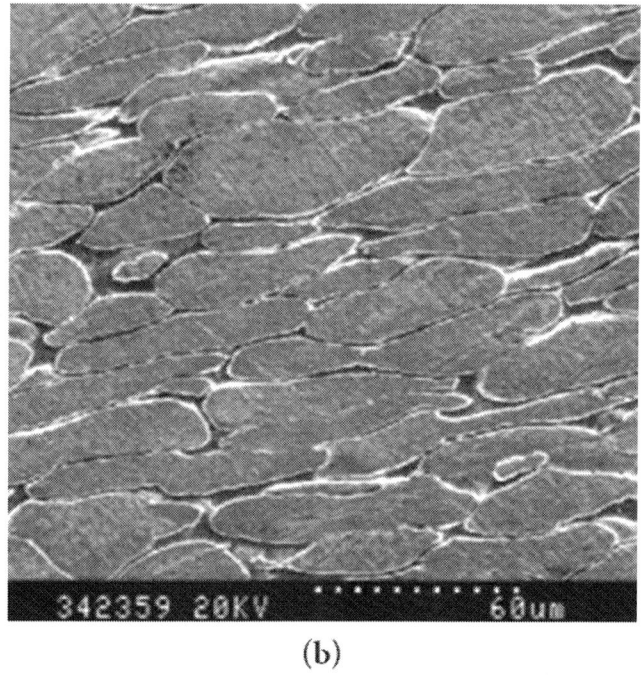

(b)

Figure 3: Longitudinal microstructure of 93W-4.9Ni-2.1Fe alloy samples: (a) as-sintered and (b) hot hydrostatically extruded at an extrusion ratio of 2.86 : 1, respectively [92].

Microwave Sintering Method

Microwave (MW) heating is a process in which the materials couple with microwaves, absorb the electromagnetic energy volumetrically, and transform into heat [93]. Microwaves directly interact with particulates within the pressed compacts rather than being conducted into the specimen from an external heat source, thereby providing rapid volumetric heating [94]. The sintering behavior of 92.5W-6.4Ni-1.1Fe compacts consolidated in MW furnace and in conventional furnace (CF) indicates that no intermetallic phase formation occurs during MW processing [95]. MW sintered W-Ni-Fe microstructure shows significantly lower coarsening of W grains. The effect of heating rate on MW sintered 90W-7Ni-3Fe heavy alloys exhibits excellent properties with the relative density over 99% [96]. With increasing the heating rate, W grain size decreases from 13.6 to 9.6 μm while the W-W contiguity increases from 0.25 to 0.35. The heating rate of 80°C/min results in the best combination of microstructure and mechanical performance. Sintering under optimized conditions leads to enhanced mechanical properties, namely, 1020 MPa tensile strength and 21% elongation.

The sintering behavior of 90W-7Ni-3Fe alloy compared with prealloyed 90W-PA (Ni-Cu) alloy in both CF and MW furnace, at various temperatures [97], shows that MW sintering requires about 80% less processing time than required by CF method. Additionally, MW sintering provides relatively fine microstructure and better mechanical properties and consumes much less energy than CF sintering. The microstructures of the differently sintered alloys are compared in Figure 4[97]. Despite such fast heating rate, no micro- or macrocracking nor distortion is observed in MW-sintered samples.

(a)

(b)

Figure 4: SEM micrographs of (a) CS and (b) MW-sintered 92.5W-6.4Ni-1.1Fe alloy [97].

A comparison on mechanical properties of CF and MW sintered alloys shows that the CF method gives some advantages like average grain size, tensile strength, and hardness [97, 98]. Average grain sizes are about 17.3 and 9.4 µm for CF and MW samples. Tensile strengths for CF and MW are about 642, 805 MPa while hardness is about 210, 295 HV for CF and MW applications, respectively. It is observed that better grain distribution and lack of intermetallic phase formation result in better mechanical properties in MW sintered compacts [99, 100].

Spark Plasma Sintering

Spark plasma sintering (SPS) is a newly developed process, synthesis, and processing technique at low temperatures and short periods. The SPS process involves the sintering of powders under the simultaneous influence of current and pressure. In this process, powders are placed in a die (typically graphite). Heating is affected by passing a current (typically pulsed DC) through the die and the sample while a pressure is applied on the powder. The characteristics, therefore, include high heating rate, application of pressure, and effect of current [101]. Higher densification is obtained since powders are sintered under an applied pressure. The SPS process is schematically shown in Figure 5 [101]. It simultaneously applies an electric current along with mechanical pressure in order to consolidate powders or synthesize and simultaneously densify specific products.

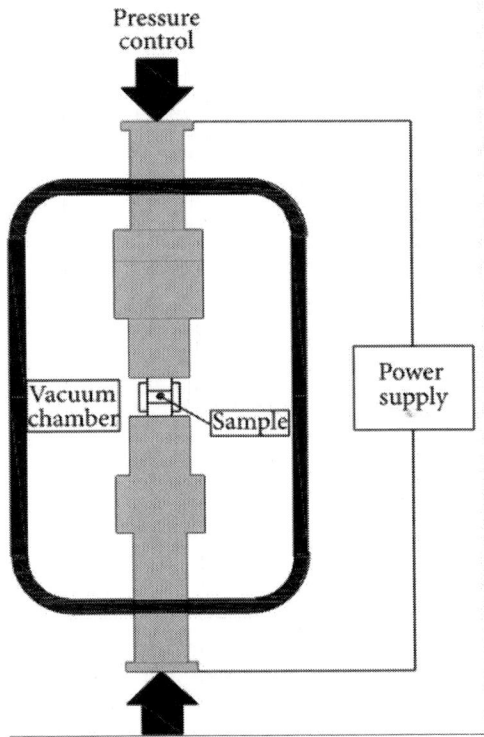

Figure 5: Schematic of the SPS process [101].

However, the SPS technology has been successfully used to prepare WHAs within limited numbers [102–108]. Almost full densification is attained at 1250°C to produce fine-grained 90W-7Ni-3Fe heavy alloys with W grain sizes of less than 5 µm. During heating, the hardness of the alloy tends to first rise, reaching the maximum value of 72.2 HRA at 1100°C, and then goes down. The microstructure of the specimen after high energy MA and pulse plasma sintering (PPS) tends to form a structure with an average grain size of ~500 nm at a sintering temperature of 1100°C [105–107]. The maximum mechanical properties up to 2500 MPa for nano- and ultradispersed W-Ni-Fe (WNF) and W-Ni-Fe-Co (WNFC) alloys are obtained. However, the PPS at high temperature above 1050°C deteriorates the strength characteristics of the alloy. For the SPS alloyed specimen, oxide film elimination of powders may improve the binding strength of the interface [108]. According to the literature [109], by optimizing the sintering times of pulsed- and constant-currents on the milled W-4Ni-2Co-1Fe powders, the density, hardness, and transverse rupture strength of the sintered alloy reach 16.78 g/cm³, HRA 84.3, and 968 MPa, respectively. Meanwhile, W grain growth in sintering is effectively inhibited, and full density is obtained at approximately 1230°C with 5 min holding time [110, 111]. Moreover, the grain growth by the SPS method can be essentially prevented [112].

MICROSTRUCTURES

Microstructure is a signature of the material processing. The microstructure is characterized by variants like grain size, shape, distributions of each phase, interface between the solid grains and matrix, porosity, and pore size in addition to processing factors such as temperature, heating rate, and holding time.

Interfacial Characteristics

To obtain composites withthe desired microstructures and properties, the interfacialreaction should be controlled through selecting an appropriatematrix alloy and controlling the process parameters liketemperature and holding time correctly.Wettability is relatedto surface energies of the interacting species by Young'sequation.The contact angle, θ, is associated with the balanceof three interfacial energies, γ_{SV}, γ_{SL}, and γ_{LV}, as follows:

$$\cos\theta = \frac{\gamma_{SV} - \gamma_{SL}}{\gamma_{LV}},$$

(1)

where the subscripts S, L, and V represent solid, liquid, and vapor, respectively. The contact angle is altered by factors that change solubility or surface chemistry. For example, the addition of Mo to TiC-Ni system decreases the contact angle from 30° to 0° [113]. The gradual variation of Mo content results in gradients of interfacial tension [114]. A low-contact angle induces liquid spreading over the solid grains, providing a capillary attraction that helps densify the system while a high contact angle indicates poor wetting, so the liquid retreats from the solid. This results in compact swelling and liquid extruding from pores. XRD observation reveals that bcc solid solution W phase, Fe_7W_6 intermetallic phase, and Fe-Ni solid solution phase form after 1 h at different temperatures.

Grain Size

Grain size is usually reported as the number of grains per unit area. The yield strength of the alloy is related to the inverse square root of its grain size, as shown by the Hall-Petch equation:

$$\sigma_y = \sigma_o + Kd^{-1/2}, \qquad (2)$$

where σ_y is the yield stress, σo is the yield strength of a single crystal, K is a constant, and d is the grain size. Figure 6 shows that the yield strengths of WHAs are well represented as a function of $(1 - V_M/DV_M)^{1/2}$ [78, 79]. These alloys exhibit a direct Hall-Petch relation above a crystallite size of 12 nm but an inverse Hall-Petch relation below 12 nm [43, 83].

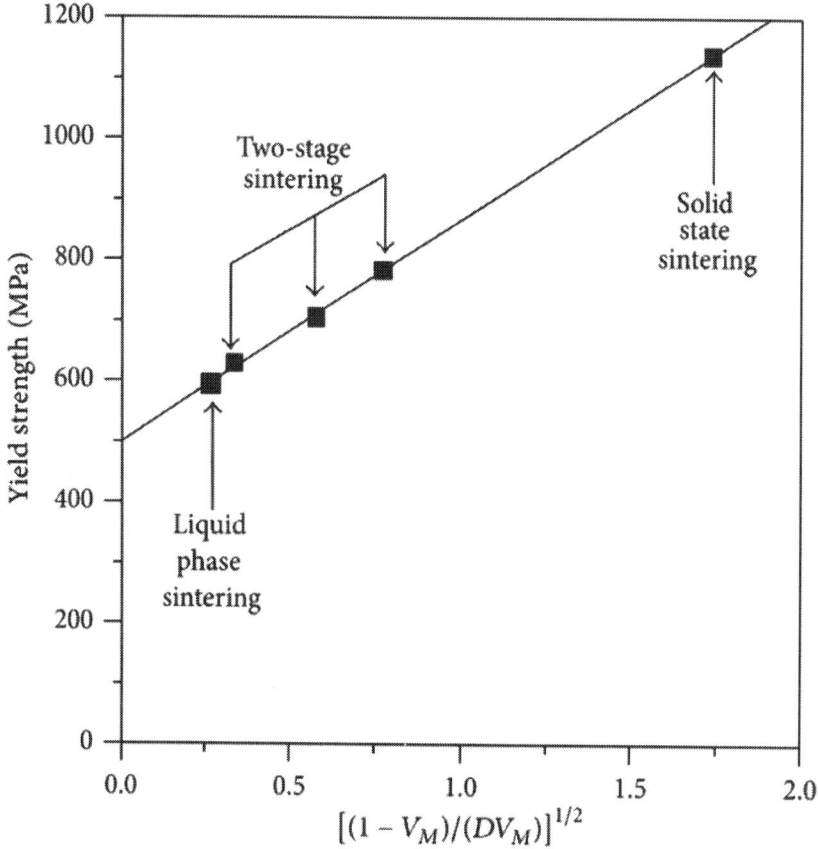

Figure 6: Variation of yield strength as a function of microstructural parameters including the matrix volume fraction, V_M, and tungsten particle size D, of solid-state sintered, two-stage sintered, and liquid-phase sintered 93W-5.6Ni-1.4Fe WHAs [78].

The grain size of crystalline W in powders decreases with increasing the milling times [81, 87, 111,115]. Increasing the milling times, amorphous transition takes place rapidly and the volume fraction of amorphous phase increases significantly [29]. Figure 7 shows the grain size of the MAed powders with increasing the milling time [82]. MA leads to a fast decrease of the grain size to less than about 11 nm after extended milling of 60 h. In addition, the residual crystalline W decreases with the milling time.

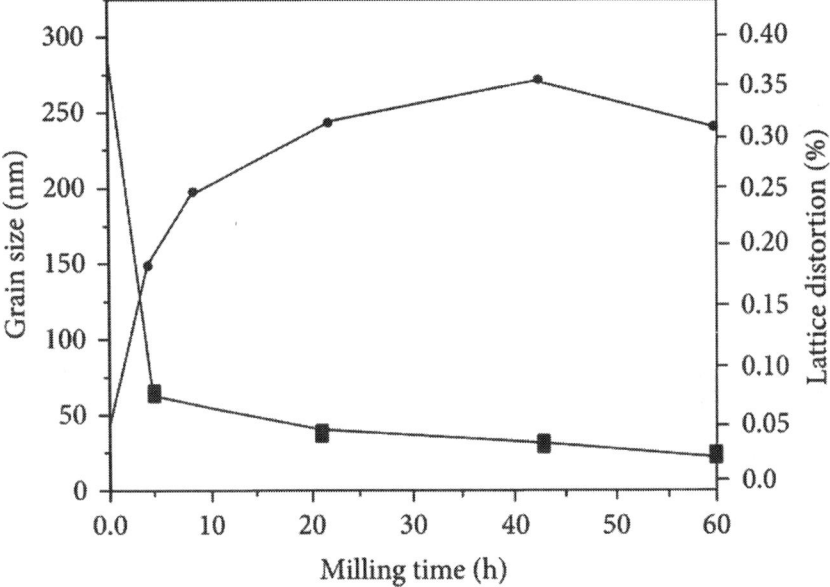

Figure 7: Variation of grain size and lattice distortion of tungsten versus milling time [82].

The isothermal grain growth kinetics during LPS follows a simple power law:

$$G^n - G_0^n = Kt, \qquad (3)$$

where G_0 is the initial grain size, t is the mean grain size, G is the sintering time, and K is the grain growth rate constant, which is related to the transport mechanism [23]. The mean grain volume increases linearly with time whereas the number of grains decreases with inverse time. The growth rate of W grain size drastically increases with annealing temperature higher than 1150°C [82]. The adding of Re and Mo prevents W atoms from solution-diffusion and inhibits W grains from precipitation growth [78, 116, 117]. Nearly full density of

90W-4Ni-6Mn alloys with sintering temperature in the 1100–1200 °C range and ultimate tensile strength (UTS) of 1000 MPa with a 25% elongation are achieved [118, 119]. The capacity of atomic diffusion can be markedly improved by the thermal activation [14]. In most of WHAs systems, the solids content exceeds 60 vol. percentages.

Porosity and Pore Size

Pores are initially present as interparticle voids, but they might also arise from inhomogeneous particle packing or volatile phases in the green body like entrapment of gases, hydrogen production, and holding time. Porosity and inclusions are detrimental to the mechanical properties of heavy alloys. If the porosity is over 0.5%, the properties of the materials, especially ductility, remarkably reduce [118–123]. As the smaller pores are filled, the mean pore size increases while the number of pores decreases. Compact swelling due to pore formation at prior particle sites is observed if the liquid forming particles have substantial solubility in the solid during heating [124, 125]. However, swelling is reduced by the use of small melt-forming particles with a similar size as the interparticle voids. The reason for the swelling or bubbles is related to impurities and oxygen led by MA process. Another approach is to use a dry-wet hydrogen combination along with Ar to avoid gas pores and hydrogen-embrittlement [83, 126]. Since the grain size increases with sintering time, liquid filling of larger pores takes considerable time.

As for the case of the SPS, neck growth enhances by applying pressure and temperature [127–129]. The material plasticity increases substantially with decreasing the W content and the deformation capability being determined by the condition of the W-W and W-matrix zone interface [130]. As the solid content increases, a rapid increase in grain size is observed [47], but the larger grain size clearly indicates more rapid grain growth. The higher the contiguity with weak cohesive strength between the contacting grains, the poorer the mechanical properties [131]. The contiguity increases with increasing the heating rate because it causes more contacts between grains [122]. The contiguity decreases as the volume fraction of liquid phase increases.

MECHANICAL PROPERTIES

Strength and stiffness are the two most important characteristics among the mechanical properties. Furthermore, yield strength, hardness, creep resistance, fatigue, and even wear resistance are decisive for structural applications. These properties can be improved or deteriorated with many factors such as alloying, composition, heat treatment, adding reinforcement, methods, production parameters, and environment. When considering WHAs, the most effective factors determining the properties are individual constituents, volume fraction of reinforcement and matrix phase, distributions of solid particles, particle size and shape, interface bonding, and microstructural factors [127–134].

Tensile Strength and Hardness

Mechanical properties are essentially functions of manufacturing processes. Both tensile hardness and elastic modulus increase with introducing the reinforcement phase and its distribution, grain size, volume fraction, matrix content, heat treatment, and production type [129]. Particle strengthening, work hardening, load transfer from particle to matrix, grain size, and uniform distributions are the key strengthening mechanisms in WHAs. For example, the dispersion of hard and fine particles in the heavy alloy drastically blocks the motion of dislocations and therefore strengthens the matrix material. Work hardening also occurs when the composite strained since the higher density of dislocation in the matrix usually took place due to the strain mismatch between the particle and matrix, thus strengthening the material [130]. In addition, load transfer is very important strengthening mechanism. The applied stress can be transferred from the soft matrix to the particle phase if the interface bonding between the hard particle and matrix is strong enough. The room-temperature tensile properties of the as-swaged tungsten alloy samples show that the ultimate tensile strength of the alloy increases significantly (Figure 8). The ultimate tensile strength increases from 750 MPa for sintered to 1640 MPa for 75% swaged alloy [90].

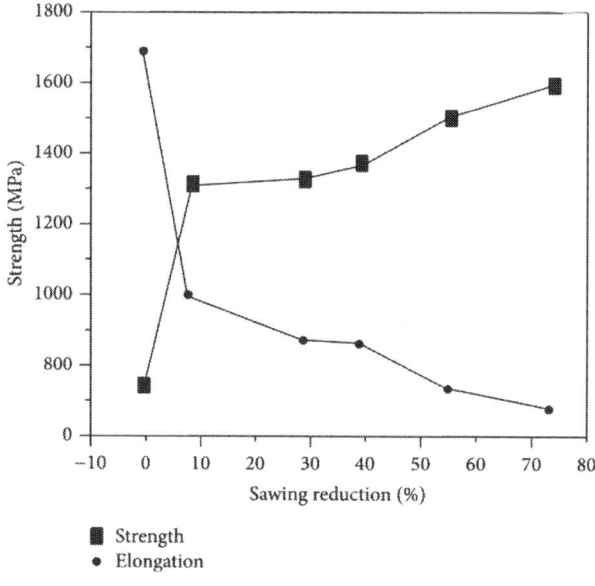

Figure 8: Variation of the strength and elongation with the swaging reduction [90].

The percent elongation to failure decreases from 20% to 6% that again can be attributed to strain hardening of both W and matrix phases. A fine-grained 90W-7Ni-3Fe alloy with high tensile strength and elongation properties is obtained, with the maximum tensile strength of 1050 MPa and elongation of 30%. In addition, the hardness for 93W-4.9Ni-2.1Fe and 93W-4.9Ni-2.1Fe-0.03Y compact significantly increases with the increase in the temperature [36, 78, 83]. The maximum hardness increased up to HV 439 ± 25 for 93W-4.9Ni-2.1Fe alloy and HV 482 ± 5 for 93W-4.9Ni-2.1Fe-0.03Y alloy after sintering at 1410°C for 1 h. Table 1 shows some mechanical and physical properties of WHAs after sintering, extruding, and heat treatment. This table shows that sintering temperature, time, composition and alloying element, milling time, method applied, and environment are the main influencing factors on microstructure and mechanical properties of heavy alloys. Among these parameters, with increasing the milling time, its grain size decreases considerably. Thus, it leads to the improvement of the tensile and hardness of these samples at room temperature. Furthermore, novel production techniques like SPS extrusion provide some advantages over the other methods. The SPS method is found to have higher yield strength than the others. The higher tensile strength is combined with approximately 24% elongation. However, extrusion method also enhances the mechanical properties significantly because of the high aspect ratio of the elongated W particles during this process, followed by two-stage MA sintered samples.

Table 1: Some mechanical and physical properties of WHAs after sintering or heat treatment obtained from the literature

Composition (wt.%)	Process	Relative density (g/cm3)	Grain size (µm)	Tensile strength or compr. strength (MPa)	Elong. (%)	Hardness (HRC/HV)	Reference
90W-7Ni-3Fe (1);	As-sintered (1);	99.7	—	890	25	27	[6]
90W-7Ni-3Fe (2);	As-extruded (2);	99.7		1260	10	36	
93W-4.9Ni-2.1Fe (3);	As-sintered (3);	—		935	21	28	
93W-4.9Ni-2.1Fe (4)	As-extruded (4)	—		1390	7	39	
90.5W-7.2Ni-1.8Fe-0.5Co-0.05Mo (1);	As-sintered (1);	99.4	25–50	950	21	33	[35]
92.5W-5Ni-2.5Cu (2);	As-sintered (2);	99.4	30	681	6	30	
93.5W-4.0Ni-2Fe-0.5Co (3)	As-sintered (3)	99.4	30	853	14	31	
95W-3.5Ni-1.5Cu (1);	As-sintered (1);	98.4	60	660	3	450 HV	[36]
91W-7Ni-1.5Fe-0.5Co (2)	As-sintered (2)	99.4	30	1000	20	433 HV	
93W-4.9Ni-2.1Fe-0.03Y	As-sintered-fine (1);	—	10	995	24	29	[36]
	As-extruded (3.33:1) (2);		10	1570	6.5	48	
	As-sintered (3);		45	960	23	29	
95W-3.5Ni-1.5Fe	As-sintered at 30min (1);	—	18	930	18.2	—	[41]
	As-sintered at 90min (2);		30	940	22		
	As-sintered at 240min (3);		41	920	17.9		
	As sintered at 640min (4)		45	836	14		
97W-2.1Ni-0.9Fe	As-sintered	99.4	19	872	21.6	—	[65]

88W-8.4Ni-3.6Fe-UDP;	As-sintered-MA (1);	97.0	5	894	30	—		[79]
93W-4.9Ni-2.1Fe;	As-sintered (2);	—	—	996	23	—		
93W-(6 –) (Ni, Fe) XPSZ, Ni/Fe = 4, X = 0.05	As-sintered and MA at 300 MPa (3)	99	22	340–425 (800°C comp. YS)	—	—		
93W-4.9Ni-2.1Fe (1);	As-sintered at 1300°C for 60 min (1);	99.6	38	—	—	438 HV		[88]
93W-4.9Ni-2.1Fe-0.03Y (2);	As-sintered at 1410°C for 60 min (2)	98.0	6	—	—	489 HV		
94W-4.8Ni-1.2Fe (3);		99.0	22	920	22			
94W-4.7Ni-1.18Fe-0.1PSZ (4); 94W-4.56Ni-1.14Fe-0.3PSZ			—	925	9			
93W-4.9Ni-2.1Fe	As-sintered (1);	99.7	4	960	23	28		[92]
	As-sawed (2.5:1);		—	1300	15.2	42,43		
	As-sawed (4:1)		—	1540	9.2	48,49		
92.5W-6.4Ni-1.1Fe	Conventional heating (1);	98.5	17	642	3.5	210/398;		[97]
	MW heating (2)	98.5	9.4	805	11.2	295/407		
90W-7Ni-3Fe	Conventional heating (1);	5°/min;	—	862	19.7	—		[98]
	Conventional heating (2);	10°/min;	20	817	12.8			
	MW heating (3);	45°/min;	—	898	20			
	MW heating (4)	105°/min	—	850	21			
95W-3.5Ni-1.5Fe-UDP at 200–1500°C (1);	Free sintering (1);	98	5–10	1300 (YS)	23 > 10	—		[104]
PPS in 900–1300°C (2);	PPS sintering (2),	98.5	—	1900 (YS)	settle			
Vacuum (3)	vacuum (3)	—	—	2500 (YS)	<0.2 s			

Recent Progress in Processing of Tungsten Heavy Alloys

90W-4Ni-6Mn (1);	As-sintered (1);	99.7	8	1030–1062	25	32	[105]
93W-5.6Ni-1.4Fe (2);	Blended-SPS (2);	95	6	600–800 (YS)	—	—	
93W-5.6Ni-1.4Fe (3);	MA/SPS, 1300°C, 1 h/1470°C, 4–90 min heat-treated, 1150°C, WQ (3);	99	6–27	716 (TS)	—	—	
91W-7Ni-1.5Fe-0.5Co (4)	CIPed, 250 MPa, 1100°C followed by WQ (4)	99.4	30	—	—	—	
95W-5 (Ni/Fe = 7/3)	As-sintered condition	—	—	991	14	—	[115]
90W-6Ni-2Fe-2Co	As-sintered-1450°C, 2 h (1);	99.6	6	450	5	430 HV	[122]
	1150°C/5 h (2);		—	750	18	435	
	W1-10% (3);		—	1348	10	490	
	W3-40% (4);		—	1430	9	530	
	W5-75% (5);		—	1646	6	610	
	W4-700°C (6);			1450	7	520	
	W4-1100°C (7)			1025	5	495	
90W-7Ni-3Fe	As-sintered at 1300–1400°C (MA)	99.7	3–1.7	1100	Very low	34	[131]
88W-8.4Ni-3.6Fe (1);	As-sintered (1);	99.00	3	894	30	—	[132]
93W-4.9Ni-2.1Fe (2);	As-sintered (2);	—	—	900	23	—	
90W-7Ni-3Fe (3)	As-sintered (3)	99.6		1000	30	—	
90W-7Ni-3Fe-0.04Y2O3	As-sintered at 1480°C for 30 min	—	99.3	1050	30.8	—	[133]
93W-4.9Ni-2.1Fe	As-sintered at 1530°C (1);	—		95	960	23	29	[134]
	As-extruded (4:1) (2)				1540	9.2	48	

Improvements of the tensile strengths in SPS and extrusions are about 51% and 40.6%, respectively. However, this reaches up to 47% when the extrusion ratio increases to 75%. For the MW heating, 13% improvements are achieved when compares to conventional one. In the case of conventional sintered samples associated with different compositions and temperature/times, 38% improvements are obtained for the tensile strength while the elongation is lower than 21%. As a result, development of method is found to be more effective on improvements of mechanical properties of WHAs than those of other parameters.

Fracture Surfaces

It is observed that there are four possible fracture paths for WHAs microstructure: matrix failure, W cleavage, W-W intergranular failure, and W-matrix interfacial separation [131]. The most fractured interfaces are composed of W-W interface boundaries in solid-state sintered 93W-5.6Ni-1.4Fe WHA and two-stage sintered 93W-5.6Ni-1.4Fe WHA [115]. Failure under tension starts by separation of W-W interface areas and develops by producing cleaved W grains after strain hardening of the matrix phase. However, at the same time, the ductility of the alloy decreases as W content increases, and above 93% W the ductility of the alloys decreases drastically. The strength of WHAs increases as the fracture surface changes from flat facets, river like, to sheared W grains [90]. Lower strength and ductility in 90W-6Ni-2Fe-0.5Co-1.5Mo alloy are attributed to the presence of a higher fraction of flat faceted W grains. Oxides at W-W and W-matrix interface severely promote fracture and thus elongation decreases [80, 83]. However, the fracture surface by impact test reveals a change of the W-W interface failure mechanism towards ductile failure of penetrated matrix with many dimples by increasing the matrix penetration [43, 50].

The microstructure and processing factors of the heavy alloys are the key to affect their mechanical properties. For WHAs, the main influencing factors of mechanical properties include compositions [27–33, 36–39, 83], processing type [71–73, 85, 95, 99, 103, 122, 132, 133] and temperature/time [41, 43, 62, 76, 86, 97, 104, 105, 134], tungsten grain size and shape effect [47, 81, 87, 118], grain growth [46, 82, 110, 112, 117], interface bonding between W and matrix [80, 90, 114, 130, 131], and porosity and pore size distribution [49, 83, 118, 124, 125, 127].

Penetrator Performance

In order to increase the penetration performance it is desirable to promote adiabatic shear banding localized to small areas and at higher shear strains. For WHAs, first, W/W interfacial area without altering W content should be reduced. Secondly, the W grain size should be increased and thus contiguity decreases. Thirdly, the coated powder should be used. Table 2 indicates requirements of an ideal KE penetrator. Among all types of penetrator materials, there are two key factors affecting the penetration ability, which are good penetration and high density [135].

Table 2: Requirements of an ideal KE penetrator

Number	Attribute	Function
1	High density	Imparts high impact energy
2	High strength	For greater heat generation for a given strain
3	Low heat capacity	Heat generated results in rapid temperature rise
4	Low work hardening rates	Easier flow softening
5	Low strain rate hardening rate	Shear localization occurs at lower strain
6	High thermal-softening rate	Shear banding initiated at lower temperature

Traditionally, depleted uranium (DU) alloys U-3/4Ti are used as penetrators in all cannon-fired KE projectiles. However, due to environmental issue and health hazardous of DU, there is a drive to replace DU alloys by alternative materials. The development of a large mushroomed head on the penetrator is observed for pure tungsten, WHAs, and a uranium-6% niobium alloy is shown in Figure 9 [135, 136].

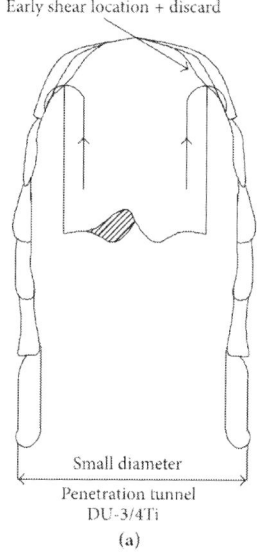
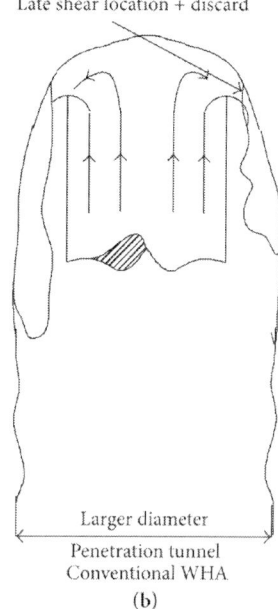

Figure 9: Penetrator deformation behavior of (a) DU rod and (b) WHA rod [135].

However, other U-3/4Ti alloy penetrators, which rapidly develop localized adiabatic shear failure, create smaller diameter penetrator cavities than WHAs penetrators. This type of failure results in a chiseled head formation (self-sharpening effect), which improves KE penetrators performance. As stated above, the former indicates self-sharpening effect, whereas the latter shows a mushrooming effect. This is the fact that the very high melting of W entails a low thermal softening associated with high rate sensitivity of W with the strength increasing with an increasing strain-rate. These effects delay flow softening at high strain rates. Because of delaying onset of shear stress localization, WHAs develop "mushroom heads" during penetration.

WHAs such as the one studied here have been studied extensively, primarily as potential materials for KE projectiles [136]. Mechanical properties and physical phenomena at high strain rates have been obtained or observed using Hopkinson bar techniques [137], normal plate impact [138–140], and oblique flyer-plate impact [141]. In most of these, the bulk mechanical properties of the homogenized multiphase material (e.g., stress-strain curves and spall strengths) and microstructure such as fracture surfaces or shear localization initiation sites have been reported.

The dynamic deformation and fracture behavior of an oxide (0.1 wt.% Y_2O_3) dispersed WHAs fabricated by MA show that interfacial bonding between W particles occurs over broad deformed areas in this alloy, suggesting the possibility of adiabatic shear banding (ASB) formation [18]. Combined dynamic compression and shear tests are carried out to study the deformation and failure behavior of pretwisted WHA at high strain rates of $10^3 s^{-1}$ [54]. It is found that the initial microstructures such as the aspect ratio and orientation of the W grains have significant effect on the tendency to the shear band formation. Presumably the ASB initiates at one side, where the direction of the maximum shear stress is approximately parallel to the direction of the major axis of W grains, and then develops at the other side due to the loss of load-bearing capacity. The ASB has general trends to propagate along the Ni-Fe matrix because more energy is needed to shear the W grains. The influence of the microstructure such as heterogeneity, the content of matrix, and the particle shape on shear band formation is reported [29, 142]. The quasistatic and dynamic failure of ultrafine grained (UFG ~ 500 nm) tungsten (W) under uniaxial compression is investigated under uniaxial dynamic compression (strain rate ~ $10^3 s^{-1}$)

[143]. The true stress-true strain curves of the UFG-W exhibit significant flow softening, and the peak stress is ~3 GPa. The strain rate sensitivity of the UFG-W is reduced to half the value of the conventional W. Furthermore, WHAs rod penetrators processed by hydrostatic extrusion and hot torsion (HE + HT) subjected to ballistic impact maintain an acute shape and show good self-sharping ability, while mushroom-like heads are observed in the as-sintered and as-extruded WHA penetrators [93]. Microstructure analysis shows that ASBs form at the edge of the HE + HT WHA penetrator heads during the penetration and deformed parts of penetrator fall along the ASBs, which are responsible for the good self-sharpening ability and evidently improved the penetration performance (see Figures 10 and 11).

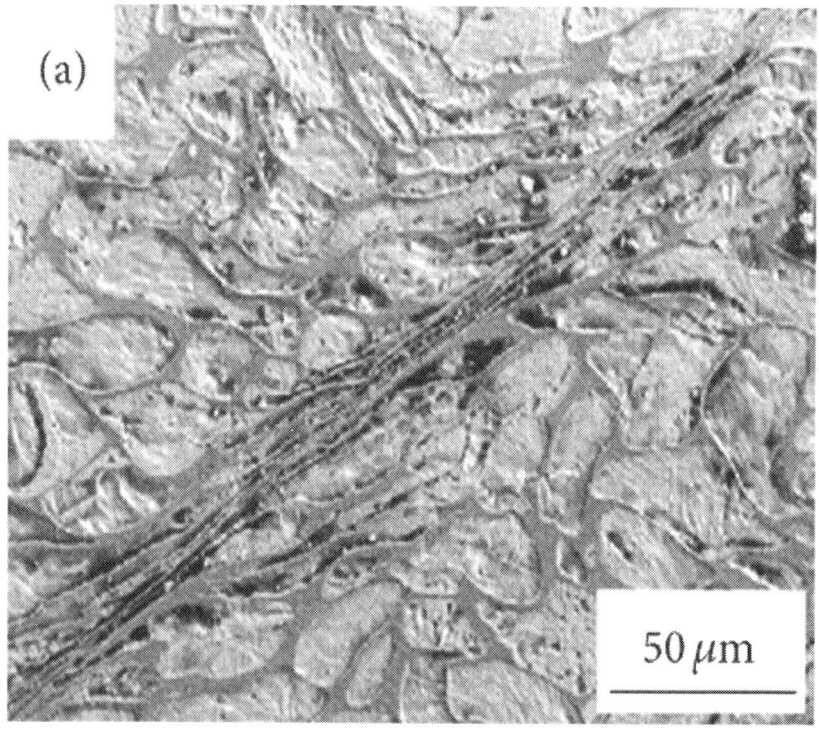

(a)

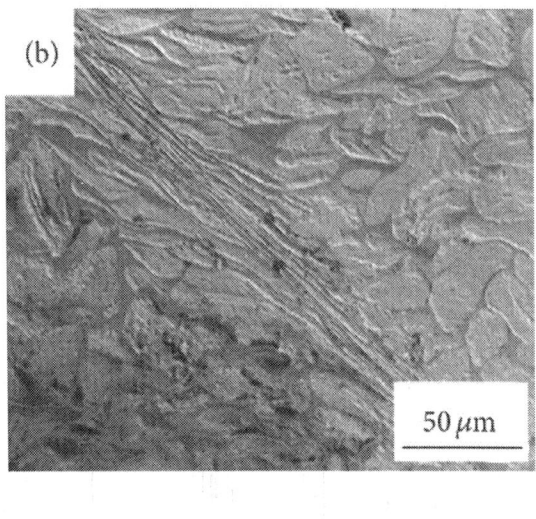

(b)

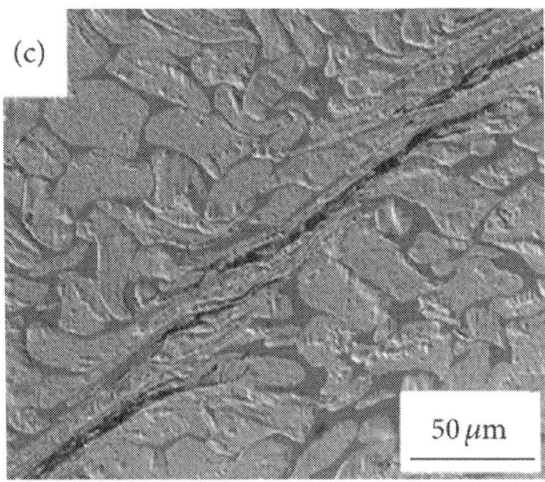

(c)

Figure 10: SEM micrographs showing ASB of specimens subjected to dynamic compression at initial temperatures of (a) 26°C, (b) −20°C, and (c) −50°C, respectively [55

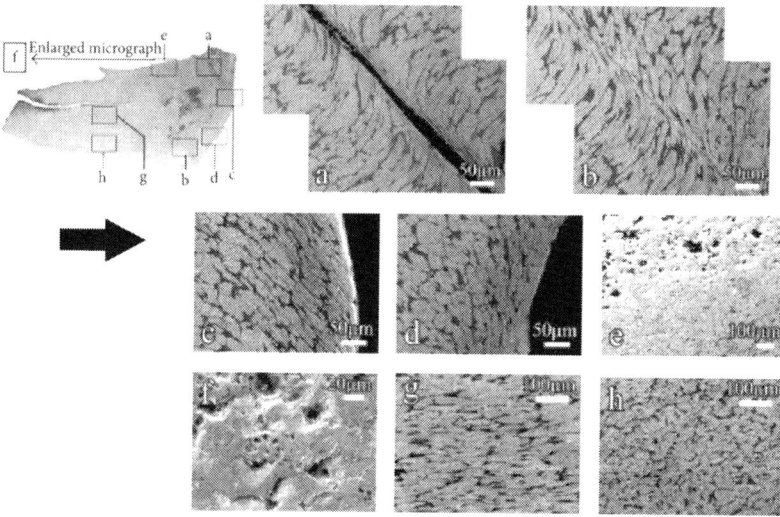

Figure 11: SEM micrographs in different regions of the HE + HT WHA penetrator remnant. The arrow depicts penetration direction. (a) Cracks formed along the ASB; (b) obvious localized shearing and microcracks within the ASB; (c) and (d) localized shear bands adjacent to the fracture face indicating that fracture happened along the ASBs; (e) large amounts of holes on one side of the remnant showing micrographs of the central region and the edge field near the tail of the remnant, respectively [93].

Deformation behavior of W-Ni-Fe alloy by impact loading over a practical range of temperatures ranging from 25 to 1100°C and strain rates ranging from 8×10^2 to $4 \times 10^3 \, s^{-1}$ has been studied by Lee et al. [144]. The flow stress of the composites increases with the strain rate, but an increase in the test temperature has opposite effects. Moreover, work hardening decreases considerably as both the temperature and the strain rate increase. However, a decrease in strain rate sensitivity is observed when the temperature goes above 700°C. Tensile impact tests performed on a tungsten heavy alloy at strain rates ranging from 100 to $1300 \, s^{-1}$ indicate that strain softening is evidenced [145]. Previous work on pressure-shear plate effects to simple shear under shearing strain rates up to $7 \times 10^5 \, s^{-1}$ shows that alloys exhibit significant rate sensitivity and thermal softening due to plastic dissipation. Shear bands form when the plastic strains become sufficiently large [26, 146]. Uranium alloys are better penetrators due to the use of strain onset for shear banding than tungsten alloys, but the assumption that

ASB plays a major role in the erosion process of WHAs remains to be verified [147, 148].

Dynamic compression failure testing against steel targets of hardness ranging from 303 to 330 hardness Brinell (HB) is conducted with an impact symmetric Taylor test allowing the quantification of ABS as a function of the impact velocity [53]. The equivalent failure strains for ABS are identified through numerical simulations of the Taylor tests with a finite element hydrocode implemented with Johnson-Cook constitutive models [25]. These data along with the chiseled nose observations are the first evidences that shear banding is one of the failure processes influencing the penetration performance of conventional WHAs penetrators. The reverse ballistic impact tests on WHAs rods are conducted at a speed of $173\,m\cdot s^{-1}$ and $228\,m\cdot s^{-1}$ by Dick et al. [149]. The mushroomed end has a slight ellipticity which attribute to radial cracks in the body at low speed while the deformation is more localized along a curved path extending from the mantle near the transition between the mushroomed region and the relatively undeformed rod toward the impact face. A cusp formed in the lateral boundary where the shear band intersects it. Opposite to a point where a shear band is formed, they observe the ductile fracture along a similar path, but no data is obtained for materials properties.

For WHAs, ASB is defined as narrow regions of rapid intense plastic deformation, forming near the transition between the mushroomed region and the relatively undeformed portion of the rod, when its material is either a mixture of Fe-N-W and W particles or a homogenous alloy [55, 150]. The ASB is observed under all specimens subjected to various initial temperatures. Figures 10(a), 10(b), and 10(c) show the SEM micrographs of ASB within specimens subjected to 26°C, −20°C, and −50°C, respectively. It can be observed that the as-extruded WHAs exhibit a good susceptibility to ASB even subjected to cryogenic temperatures. Furthermore, with the decrease of initial temperature, the shear strain within ASB shows a decreasing tendency, demonstrating that the initial temperature has some influences on adiabatic shear banding of WHAs. The shear banding in a liquid phase sintered 90W-7Ni-3Fe alloy at $5400\,s^{-1}$ appeared [129]. Several W-Ni-Fe alloys using compressing Kolsky test at high strain rates for dynamic deformations involving large plastic strains, the response of the WHAs, are controlled by the behavior of W grains [137]. The overall mechanical properties can be improved when the W particles are of fine size and the interfacial area between W particles is reduced [104]. However, penetrators that have too high strength are easily embrittled, and, thus, their penetration performance deteriorates because of fragmentation during penetration. It is desirable for a penetrator to have deformation behavior that can be induced self-sharpening and possess sufficient fracture toughness to minimize fragmentation during high-speed impact [142]. A shear band forms at the point on the rods mantle where the mushroomed region transitions into the straight rod, which agreed qualitatively with Dick et al. [149]. However, the rod is made of either homogeneous tungsten or homogeneous Fe-Ni-W and no such shear band is computed.

Thermomechanical deformations of a WHA rod impacting at normal incidence of a smooth rigid target are modeled based on Fe-Ni-W particles randomly distributed in W particles, homogenous WHAs, pure tungsten, and pure Fe-Ni-W [150]. The role of microstructure on dynamic mechanical properties of WHAs has been investigated via constitutive modelling and numerical simulation. An isotropic elastoviscoplastic constitutive model with realistic grain structures to capture shear location in oblique impact (pressure-shear conditions) has used [26] and shows high rate nominally uniform shear deformation [151].

These deformation models and their applications for WHAs rods will be described in the following section.

Formulation

In order to describe dynamic, thermomechanical, and axisymmetric deformations of a cylindrical rod affecting at normal incidence a rigid and planar surface, the Lagrangian or referential description of motion is used. The deformations of rod materials are governed by the following balance laws of mass, linear momentum, moment of momentum, and internal energy.

Equation (4) is the Johnson-Cook relation [25]. The flowstress, $\sigma_{y'}$ increases with an increase in the ϵ_p effective plasticstrain and the $\dot{\epsilon}_p$ effective plastic strain rate but decreaseswith an increase in the temperature of a material particle. C is the specific heat, θ the temperature of a material particle, θm its melting temperature, θo the room temperature, T the homologous temperature, Se the effective stress (σ_y), and ϵ_p the effective plastic strain. In (4), parameters B and n characterize the strain hardening of the material, C and $\dot{\epsilon}_o$ characterize its strain-rate hardening, and $(1 - T^m)$ characterizes its thermal softening. For the Taylor impact test, initially, the cylinder rod is stress free, at room temperature θ, and is moving with a uniform speed V_o in a direction normal to the plane surface of the target and strikes it at time $t = 0$. All bounding surfaces of the rod except that contacting the target are to be taken traction free. No thermal boundary conditions are needed due to the assumption of locally adiabatic deformations:

$$S_e = \left(A + B(\epsilon_p)^n\right)$$
$$\times \left(1 + C \ln\left(\frac{(2/3) \Lambda S_e}{\dot{\epsilon}_o}\right)\right)(1 - T^m). \quad (4)$$

In physical experiments, fracture in the form of a crack will ensue from the point well before it is heated to its melting temperature. The thermal softening of the WHA they tested is better described by $(1 - (-1 + (\theta/\theta_o)^\alpha))$ where α and β are material parameters and for their WHA, $\beta = 2.4$ and $\alpha = 0.2$ [26]. Its approximate solution is by the finite element method and employs a large-scale explicit code DYNA2D [152] to do so. As the bodies deform, elements near the target/penetrator interface become distorted severely and the time step size drops drastically.

Taylor Impact Test Simulations

The Johnson-Cook Thermal Softening and Zhou et al. Thermal Softening

The following values are assigned to various material and geometric parameters.

Rod length = 60 mm; Rod diameter = 10 mm; V_o = 150 m/s; θ_o = 293 K.

Depleted Uranium (DU) [153]. $A = 1;079$ MPa; $B = 1;120$ MPa; $C = 0.007$; $n = 0.25$; $m = 1.0$; $\epsilon'_o = 1/s$; $\rho = 18;600$ kg/m³; $\mu = 58$ GPa; $K = 119$ GPa; $c = 117$ J/kg °C, $\theta_m = 1473$ K.

TungstenHeavyAlloy (WHA). $A = 1,506$ MPa; $B = 177$ MPa; $= 0.016$; $n = 0.12$; $m = 1:0$; $\epsilon'_o = 1/s$; $\rho = 18;600$ kg/m³; $\mu = 160$ GPa; $K = 328$ GPa; $c = 134$ J/kg °C; $\theta_m = 1723$ K.

The values of A, B, C, θ_m, m, ϵ'_o, and n for both materialsare taken from Rajendran's report [154] while those of othermaterial parameters are taken from a handbook. It can benoted that the shearmodulus forWHAequals 2.76 times thatfor DU. Computed results for a trial problemindicate that theheight of the mushroomed region is approximately 5 mm.

As for Zhou et al. [26] thermal softening, results for theDU andWHA rods with $\beta = 2.4$ and $\alpha = 0.2$ and 0.3 can becomputed, by keeping values of the material and geometricparameters given above. Each one of these values of α giveshigher thermal softening than that given by the Johnson-Cook relation. Thus, for same values of ϵ_p and ϵp, theflowstress will be lower with Zhou et al. softening than that withJohnson-Cook relation. It is found that $\alpha = 0.2$ and $\beta = 2.4$provided a good fit to the test data for the WHA they used.Here the same values of α for DU are considered because ofthe lack of data. Thus, the differences for their deformationpatterns will be due to the variation in the values of the shearmodulus, bulk modulus, and parameters A, B, C, and n in(4). It can be noted that the minimum value of the thermalsoftening function can be taken as zero. The height of themushroomed region is smaller for$\alpha = 0.3$and it forms soonerin each rod as compared to that for $\alpha = 0.2$ [153]. The shapesof the mushroomed regions suggest that severe deformationsand hence a shear band may initiate at one or more of thefollowing three locations: adjacent to the stagnation point(P), near the periphery of the impacted end (Q), and closeto the inflection point in the curve describing the mantle ofthe deformed rod (R). Figures 12(a), 12(b), and 12(c) showthe deformed meshes of the DU and WHA rods for each ofthe three thermal softening functions [154]. For $\alpha = 0.2$, theaforestated conditions for the initiation of a shear band aresatisfied at point R for the WHA rod and the distortions inthe deformed mesh confirm the initiation of a shear bandthere. However, for the DU rod, even though the effectiveplastic strain at point R grows rapidly first, its rate of growthtapers off.The deformedmesh shown in Figure 12(b) suggeststhat no shear band initiates from point R since the mesh isregular in the neighborhood of R. For each thermal softeningconsidered, the mesh around point P is severely deformed.In addition, the effective stress there drops to less than 80%of its peak value.Thus, according to the definition of the dropin the effective stress, a shear band initiates at point P for eachthermal softening studied.

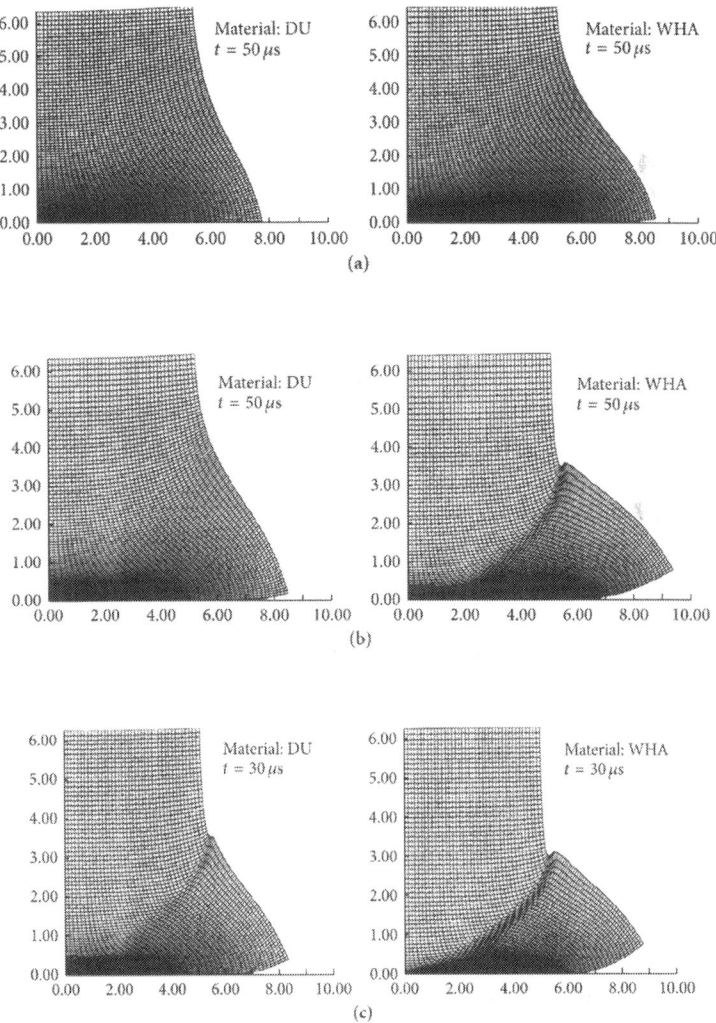

Figure 12: Deformed meshes of the impacted ends of the DU and WHA rods; (a) Johnson-Cook thermal softening, $m = 1.0$; (b) Zhou et al.thermal softening, $\alpha = 0.2$; (c) Zhou et al. thermal softening, $\alpha = 0.3$ [25, 26, 153].

Rule of Mixture Model

WHA is modeled as a mixture of 10%Fe-Ni-W particles interspersed randomly in hard tungsten particles. The two are modelled as elastic thermoviscoplastic materials with the flow stress given by (4), but with the Zhou et al. thermal softening and the following values are assigned to different materials [26, 153].

Tungsten Particles. $A = 730$ MPa; $B = 562$ MPa; $C = 0.029$; $= 0.0751$; $K = 317$ GPa; $\mu = 155$ GPa; $\beta = 2.4$; $\alpha = 0.15$; $\dot{\epsilon}_o = 1.4 \times 10^{-13}$/s; $\rho = 19300$ kg/m^3; $c = 138$ J/kg °C.

Fe-Ni-W Matrix. $A = 150$ MPa; $B = 546$ MPa; $C = 0.0838$; $= 0.208$; $K = 202.4$ GPa; $\mu = 98.84$ GPa; $\beta = 2.4$; $\alpha = 0.2$; $\dot{\epsilon}_o = 6.7 \times 10^{-14}$/s; $\rho = 9200$ kg/m^3; $c = 382$ J/kg °C.

The diameter and the length of the rod are 6.35mmand 25.4 mm, respectively, and simulations of the Taylorimpact test are carried out for two impact speeds of 173 and228m/s [154]. For $V_o = 173$ m/s, the observed deformationlengths and the geometric mean diameter of the impact faceare 22.8mm and 9.36 mm, respectively; the correspondingcomputed values equal 22.25mm and 10.72 mm. For $V_o = 228$ m/s, the computed diameter equals 9.5 mm. The finalobserved and computed deformed lengths equal 20.9mmand 20.5 mm, respectively.

In order to address the effects of crystallographic orientation and dislocation accumulation on the thermomechanical response of WHAs, crystal plasticity based constitutive models have recently been formulated and implemented by some people. The effect of variable ratios of grain and phase boundary strengths on uniaxial tensile failure shows that it is apparent that the variability in spall statistics observed experimentally is due both to the heterogeneous local strengths and flaw distributions and to the multiphase polycrystalline microstructure of the material [154]. Tensile deformation and fracture of realistically rendered [155] and homogenized the microstructures of dual phase tungsten alloys [156]. Numerical results are obtained from a 2D finite element implementation of the crystalline elastoplasticity models for each phase under impact loading. The cohesive laws account for stress state, temperature effects on interfacial strengths, and anisotropic continuum damage mechanics. They correlate with the spall strength of the multiphase material [157]. The results indicate the relationships between microstructural features, fracture patterns, and free surface velocity statistics. Statistical variations of free surface velocity are also found to vary with crystal orientation, texture, and temperature dependence of properties controlling the grain boundary fracture [158]. The role of microstructure in the dynamic deformation and fracture of a dual phase, polycrystalline tungsten alloy oriented either perpendicular or parallel to the extrusion direction under high-rate impact loading is investigated via experiments and modeling [159]. Furthermore, finite deformation crystal plasticity theory describes the behavior of the pure tungsten and binder phases, and a stress- and temperature-based cohesive zone model captures fracture at grain and phase boundaries in the microstructure. The results from both experiments and modeling indicate that the grain orientations affect the free-surface velocity profile and spall behavior.

As for the case of Spalling; dynamic tensile failure in uniaxial strain loading, typically referred to as spall or spallation, frequently occurs in impact events, including the current plate impact experiments [160]. The spall strength of a similar variety of WHA to that currently studied has been measured to be 1.7–2.0 GPa [161]. In a typical spall experiment, impact sends a shock wave or elastic and plastic waves, depending on stress level, through both the target and impactor. The initial shocked state was defined by stress and particle velocity. When the compression waves reach the back of the target, a stress-free surface, and the back of the impactor, which has a low-impedance material against it, most of the wave reflects as an unloading wave. This unloading wave from the free surface at the back of the target, also referred to as a rarefaction fan because of its tendency to spread as it propagates, unloads the sample to zero stress and the particle velocity. It can be noted that this free-surface velocity is somewhat lower than the impact velocity. Dandekar [162] attempted to determine the impedance of WHA directly through concurrently backing the target material with void and PMMA. By assuming that the same spall strength is reached in the two cases, the impedance can be found. Instead, the slope (impedance) from the shocked state to the unloaded state in the stress plane can be calculated. It is clear that the pullback velocity varies considerably with position in the sample. The average value of spall strength for the experiment is somewhat larger than the spall strength inferred from the spatially averaged particle velocity.

Experimentally, higher velocity impacts yield a somewhat higher spall strength for both orientations. Such behavior has been observed previously in other brittle materials [162]. Post mortem inspection of fully fractured (i.e., spalled) specimens would likely indicate both intergranular and cleavage mechanisms [138]. However, crack propagation and macroscopic spall behavior are thought to require substantial grain cleavage: once sufficiently large, a microcrack initiated along grain/phase boundaries will then propagate fully across the specimen, irrespective of the underlying microstructure. Effective stress near the spall plane can be simulated at different materials and impact velocities but not discussed here. A multiscale descriptive framework for describing effects of micro- or mesostructures on wave propagation in solids under impact loading has been developed [163]. The initiation of spall fracture may be controlled by grain and phase boundaries, in agreement with experimental observations.

POTENTIAL APPLICATIONS OF WHAS

WHAs emerged as the potential candidates for kinetic energy (KE) penetrators in addition to radiation shields and vibration damper applications [41, 65, 88, 131]. These alloys are excellent materials for the shielding against gamma rays and hard X-rays due to their combination of radiographic density, machinability, strength, and low toxicity. Counterbalance weights for fixed and rotary wing aircraft have been used for many years because unlike lead, which creeps under its own weight at room temperature. WHAs are routinely used in high density fragmenting devices and armor piercing ammunition ranging from small caliber of 5.56 mm diameter up to 120 mm antitank projectiles and beyond. Rotational inertial members are, for example, gyroscope rotors in inertial guidance systems and navigation equipment. WHAs serve as flywheels, the rim components of flywheels, or rotating members in governors. WHAs weight blocks offer up to 50% more weight in a given volume with added advantages of direct attachment via threaded holes or thru-bolting and the freedom from deformation, which is the constant problem for the lead in addition to hole logging of oil wells.

CONCLUDING SUMMARY

A review of recent progress on the processing such as conventional powder metallurgy (PM), mechanical alloying (MA), microwave sintering (MWS), and spark-plasma sintering (SPS) methods and microstructure and mechanical properties of WHAs is presented in this paper. Key factors affecting the performance of WHAs are given as follows. Firstly, W and matrix composition, their shape, size of W particles, distributions of W particles, and pore size are important. Secondly, improvements in strength/stiffness and hardness are observed in most alloy systems with additions of W in Fe-Ni/Cu alloys. The bonding strength between W particle and matrix interface is very important in the strengthening of alloys since the load transfer depends on the sufficient bonding between the W and matrix phase. In addition, optimal sintering temperature and time are important because of the formation of a strong interface between particles and matrix. Moreover, ultimate tensile strength and hardness of WHAs produced by extrusion method enhanced significantly because of the high aspect ratio of the elongated W particles. Thirdly, novel processing technique like SPS provides some advantages over the other methods. The SPS alloyed is found to have higher yield strength than the others, followed by extruded sintered samples. Fourthly, grain refinement is an effective way for improving ductility and strength at ambient temperature with increasing the milling time. Finally, the yield strength of WHAs depends on the microstructural parameters such as tungsten grain size, matrix volume fraction, pore size, and W-W contiguity. Furthermore, the thermal softening in DU and WHA penetrator rod are modelled by either the Johnson-Cook relation [25] and that proposed by Zhou et al. [26] or rule of mixtures. The flow stress increases with an increase in the effective plastic strain and effective plastic strain-rate but decreases with a rise in the temperature. The height of the mushroomed region is approximately 5 mm, but the height of the mushroomed region is found to be smaller for $\alpha = 0.3$ and it forms sooner in each penetrator rod as compared to that for $\alpha = 0.2$ value.

ACKNOWLEDGMENTS

The Council of Higher Education (Turkey) supports the author's research. The author also extends his thanks to Associate Professor

Alex O. Aning at the Department of Materials Engineering in Virginia Tech & State University during this study.

REFERENCES

1. J. Reiser, M. Rieth, B. Dafferner, and A. Hoffmann, "Tungsten foil laminate for structural divertor applications—basics and outlook," Journal of Nuclear Materials, vol. 423, no. 1–3, pp. 1–8, 2012. · ·
2. M. Greger, L. ížek, and M. Widomska, "Structure and mechanical properties of formed tungsten based materials," Journal of Materials Processing Technology, vol. 157-158, pp. 683–687, 2004. · ·
3. S. W. H. Yih and C. T. Wang, Tungsten, Sources, Metallurgy, Properties, and Applications, Plenum Press, New York, NY, USA, 1981.
4. E. Antolini and E. R. Gonzalez, "Tungsten-based materials for fuel cell applications," Applied Catalysis B: Environmental, vol. 96, no. 3-4, pp. 245–266, 2010. · ·
5. J. E. Emsley, The Elements, Oxford University Press, New York, NY, USA, 2nd edition, 1991.
6. X. Gong, J. L. Fan, F. Ding, M. Song, and B. Y. Huang, "Effect of tungsten content on microstructure and quasi-static tensile fracture characteristics of rapidly hot-extruded W-Ni-Fe alloys," International Journal of Refractory Metals and Hard Materials, vol. 30, no. 1, pp. 71–77, 2012. · ·
7. S. G. Caldwell, "Tungsten heavy alloys," in Powder Metal Technologies and Applications, P. W. Lee and R. Lacocca, Eds., vol. 7, pp. 914–921, ASME Handbook, Materials Park, Ohio, USA, 1998.
8. A. Upadhyaya, "Fabrication of tungsten-copper composites through liquid phase sintering,"Transactions of the Indian Institute of Metals, vol. 55, pp. 51–69, 2002.
9. G. H. S. Price, C. J. Smithells, and S. V. Williams, "Sintered alloys part IA, alloys sintered with a liquid phase present," Journal of the Institute of Metals, vol. 62, pp. 239–264, 1938.

10. A. I. Prill, "The role of phase relationships in activates sintering of tungsten," Transactions of the AIME, vol. 230, pp. 769–772, 1964.
11. N. C. Kothari, "Densification and grain growth during liquid-phase sintering of tungsten-nickel-copper alloys," Journal of The Less-Common Metals, vol. 13, no. 4, pp. 457–468, 1967. · ·
12. R. M. German, P. Suri, and S. J. Park, "Review: liquid phase sintering," Journal of Materials Science, vol. 44, no. 1, pp. 1–39, 2009. · ·
13. K. N. Ramakrishnan and G. S. Upadhayaya, "Effect of composition and sintering on thedensification and mechanical properties," Journal of Materials Science Letters, vol. 9, pp. 456–459, 1990.
14. J. F. Kuzmic, in Modern Development in Powder Metallurgy, H. H. Hausner, Ed., vol. 3, pp. 166–171, Plenum Press, New York, NY, USA, 1996.
15. V. Srikanth and G. S. Upadhyaya, "Effect of tungsten particle size on sintered properties of heavy alloys," Powder Technology, vol. 39, no. 1, pp. 61–67, 1984. · ·
16. A. Sunwoo, S. Groves, D. Goto, and H. Hopkins, "Effect of matrix alloy and cold swaging on micro-tensile properties of tungsten heavy alloys," Materials Letters, vol. 60, no. 3, pp. 321–325, 2006. · ·
17. H. J. Ryu and S. H. Hong, "Fabrication and properties of mechanically alloyed oxide-dispersed tungsten heavy alloys," Materials Science and Engineering A, vol. 363, no. 1-2, pp. 179–184, 2003. · ·
18. S. Park, D.-K. Kim, and S. Lee, "Dynamic deformation behavior of an oxide-dispersed tungsten heavy alloy fabricated by mechanical alloying," Metallurgical and Materials Transactions A: Physical Metallurgy and Materials Science, vol. 32, no. 8, pp. 2011–2020, 2001. · ·
19. J. X. Liu, S. K. Li, X. Q. Zhou, Z. H. Zhang, H. Y. Zheng, and Y. C. Wang, "Adiabatic shear banding in a tungsten heavy alloy processed by hot-hydrostatic extrusion and hot torsion,"Scripta Materialia, vol. 59, no. 12, pp. 1271–1274, 2008. ·
20. W. D. Cai, Y. Li, R. J. Dowding, F. A. Mohamed, and E. J. Lavernia, "A review of tungsten-based alloys as kinetic energy penetrator

materials," Reviews in Particulate Materials, vol. 3, pp. 71–131, 1995.

21. L. S. Magnes, in Proceedings of the 4th International Conference on Tungsten, Refractory Metals and Alloys: Processing, Properties and Applications, pp. 41–58, Metal Powder Industries Federation in Cooperation with the Refractory Metals Association and APMI International, Lake Buena Vista, Fla, USA, November 1997.

22. Z. Zhaohui and W. Fuchi, "Research on the deformation strengthening mechanism of a tungsten heavy alloy by hydrostatic extrusion," International Journal of Refractory Metals and Hard Materials, vol. 19, no. 3, pp. 177–182, 2001. · ·

23. X. Liu, X. Zhang, S. Yang, C. Zhao, and C. Wang, "Experimental investigation of phase equilibria in the Co-W-Nb ternary system," Intermetallics, vol. 31, pp. 48–54, 2012. · ·

24. R. M. German, Sintering Theory and Practice, John Wiley & Sons, New York, NY, USA, 1996.

25. G. R. Johnson and W. H. Cook, "A constitutive model and data for metals subjected to large strains, high strain rate, and high temperatures," in Proceedings of the 7th International Symposium on Ballistics, pp. 541–548, The Hague, The Netherlands, April 1983.

26. M. Zhou, A. Needleman, and R. J. Clifton, "Finite element simulations of shear localization in plate impact," Journal of the Mechanics and Physics of Solids, vol. 42, no. 3, pp. 423–458, 1994. · ·

27. P. C. Angelo and R. Subramanian, Powder Metallurgy: Science, Technology and Applications, Eastern Economy Edition, Prentice-Hall, New Delhi, India, 2008.

28. W. Liu, Y. Ma, and B. Huang, "Adiabatic shear banding in a tungsten heavy alloy processed by hot-hydrostatic extrusion and hot torsion," Bulletin of Materials Science, vol. 3, pp. 1–6, 2008.

29. A. Bose and R. M. German, "Matrix composition effects on the tensile properties of tungsten-molybdenum heavy alloys," Metallurgical Transactions A, vol. 21, no. 5, pp. 1325–1327, 1990. · ·

30. P. B. Kemp and R. M. German, "Grain growth in liquid-phase-sintered W-Mo-Ni-Fe alloys," Journal of The Less-Common Metals, vol. 175, no. 2, pp. 353–368, 1991. · ·

31. K.-H. Lin, C.-S. Hsu, and S.-T. Lin, "Variables on the precipitation of an intermetallic phase for liquid phase sintered W-Mo-Ni-Fe heavy alloys," International Journal of Refractory Metals and Hard Materials, vol. 20, no. 5-6, pp. 401–408, 2002. · ·
32. T.-Y. Chan and S.-T. Lin, "Microstructural evolution on the sintered properties of W-8 pct Mo-7 pct Ni-3 pct Fe alloy," Journal of Materials Science, vol. 35, no. 15, pp. 3759–3765, 2000. · ·
33. C. S. Hsu, P. C. Tsai, and S. T. Lin, "Segregation of molibdenim atoms to the liquid–solid interface sintering of W-8% Mo-7% Ni-3% Fe," Metallurgical and Materials Transactions A, vol. 31, pp. 1–7, 2000.
34. M. Zahrae, H. Arabi, M. T. Salehi, and M. Tamizifar, "Development of a tungsten heavy alloy, W-Ni-Mn, used as kinetic energy penetrator," Iranian Journal of Materials Science and Engineering, vol. 4, no. 3-4, pp. 9–13, 2007.
35. J. Das, U. R. Kiran, A. Chakraborty, and N. E. Prasad, "Hardness and tensile properties of tungsten based heavy alloys prepared by liquid phase sintering technique," International Journal of Refractory Metals and Hard Materials, vol. 27, no. 3, pp. 577–583, 2009. · ·
36. J. Das, G. A. Rao, and S. K. Pabi, "Microstructure and mechanical properties of tungsten heavy alloys," Materials Science and Engineering A, vol. 527, no. 29-30, pp. 7841–7847, 2010. · ·
37. U. R. Kiran, S. Venkat, B. Rishikesh, V. K. Iyer, M. Sankaranarayana, and T. K. Nandy, "Effect of tungsten content on microstructure and mechanical properties of swaged tungsten heavy alloys," Materials Science and Engineering A, vol. 582, pp. 389–396, 2013. · ·
38. X. Gong, J. L. Fan, F. Ding, M. Song, B. Y. Huang, and J. M. Tian, "Microstructure and highly enhanced mechanical properties of fine-grained tungsten heavy alloy after one-pass rapid hot extrusion," Materials Science and Engineering A, vol. 528, no. 10-11, pp. 3646–3652, 2011. · ·
39. K. R. Sriraman, S. Ganesh Sundara Raman, and S. K. Seshadri, "Synthesis and evaluation of hardness and sliding wear resistance of electrodeposited nanocrystalline Ni-Fe-W alloys,"Materials Science and Technology, vol. 22, no. 1, pp. 14–22, 2006. · ·

40. J. L. Johnson and R. M. German, "Solid-state contributions to densification during liquid-phase sintering," Metallurgical and Materials Transactions B, vol. 27, no. 6, pp. 901–909, 1996. ·
41. R. M. German, A. Bose, and S. S. Mani, "Sintering time and atmosphere influences on the microstructure and mechanical properties of tungsten heavy alloys," Metallurgical Transactions A, vol. 23, no. 1, pp. 211–219, 1992. · ·
42. Y. B. Zhu, Y. Wang, X. Y. Zhang, and G. W. Qin, "W/NiFe phase interfacial characteristics of liquid-phase sintered W-Ni-Fe alloy," International Journal of Refractory Metals and Hard Materials, vol. 25, no. 4, pp. 275–279, 2007. · ·
43. J. W. Noh, E. P. Kim, H. S. Song, W. H. Baek, K. S. Churn, and S. K. L. Kang, "Matrix penetration of w/w grain," Metallurgical Transactions A, vol. 24, no. 11, pp. 2411–2416, 1993. ·
44. Y. Wu, R. M. German, B. Marx, P. Suri, and R. Bollina, "Comparison of densification and distortion behaviors of W-Ni-Cu and W-Ni-Fe heavy alloys in liquid-phase sintering,"Journal of Materials Science, vol. 38, no. 10, pp. 2271–2281, 2003. · ·
45. P. Lu, X. Xu, W. Yi, and R. M. German, "Porosity effect on densification and shape distortion in liquid phase sintering," Materials Science and Engineering A, vol. 318, no. 1-2, pp. 111–121, 2001. · ·
46. J. H. Huang, G. A. Zhou, C. Q. Zhu, S. Q. Zhang, and H. Y. Lai, "Influence of pre-alloyed Ni-Fe-Mo binder metal on properties and microstructure of tungsten heavy alloys," Materials Letters, vol. 23, no. 1–3, pp. 47–53, 1995. · ·
47. R. M. German and A. Bose, "Fabrication of intermetallic matrix composites," Materials Science and Engineering A, vol. 107, pp. 107–116, 1989. · ·
48. D. F. Heaney and R. M. German, "New grain growth concepts in liquid phase sintering,"Advances in Powder Metallurgy and Particulate Materials, vol. 3, pp. 303–311, 1994.
49. P. Lu and R. M. German, "Multiple grain growth events in liquid phase sintering," Journal of Materials Science, vol. 36, no. 14, pp. 3385–3394, 2001. · ·
50. J. Das, G. Appa Rao, S. K. Pabi, M. Sankaranarayana, and B. Sarma, "Deformation behaviour of a newer tungsten heavy

alloy," Materials Science and Engineering A, vol. 528, no. 19-20, pp. 6235–6247, 2011. · ·

51. R. S. Coates and K. T. Ramesh, "The rate-dependent deformation of a tungsten heavy alloy," Materials Science and Engineering A, vol. 145, no. 2, pp. 159–166, 1991. · ·

52. Y. Umakoshi, T. Nakano, E. Yanagisawa, T. Takezoe, and A. Negishi, "Effect of alloying elements on anomalous strengthening of $NbSi_2$-based silicides with C40 structure," Materials Science and Engineering A, vol. 239-240, pp. 102–108, 1997.

53. H. Couque, G. Nicolas, and C. Altmayer, "Relation between shear banding and penetration characteristics of conventional tungsten alloys," International Journal of Impact Engineering, vol. 34, no. 3, pp. 412–423, 2007. · ·

54. Z. Wei, J. Yu, S. Hu, and Y. Li, "Influence of micro structure on adiabatic shear localization of pre-twisted tungsten heavy alloys," International Journal of Impact Engineering, vol. 24, no. 6-7, pp. 747–758, 2000. · ·

55. W. Guo, J. Liu, J. Yang, and S. Li, "Effect of initial temperature on dynamic recrystallization of tungsten and matrix within adiabatic shear band of tungsten heavy alloy," Materials Science and Engineering A, vol. 528, no. 19-20, pp. 6248–6252, 2011. · ·

56. D. Rittel, Z. G. Wang, and M. Merzer, "Adiabatic shear failure and dynamic stored energy of cold work," Physical Review Letters, vol. 96, no. 7, Article ID 075502, 2006. · ·

57. H. Ye, X. Y. Liu, and H. Hong, "Fabrication of metal matrix composites by metal injection molding—a review," Journal of Materials Processing Technology, vol. 200, no. 1–3, pp. 12–24, 2008. · ·

58. A. Bose, "Net shaping concepts for tungsten alloys and composites," Powder Metallurgy, vol. 46, no. 2, pp. 121–126, 2003. · ·

59. N. H. Loh, S. B. Tor, and K. A. Khor, "Production of metal matrix composite part by powder injection molding," Journal of Materials Processing Technology, vol. 108, no. 3, pp. 398–407, 2001. · ·

60. A. Bose, R. J. Dowding, and G. M. Allen, "Power injection molding of a 95W-4Ni-1Fe alloy," in Powder Injection Molding Symposium, P. H. Booker, J. Gaspervich, and R. M. German, Eds., pp. 261–274, MPIF, Princeton, NJ, USA, 1992.

61. I. H. Moon, S. S. Ryu, and J. C. Kim, "Sintering behavior of mechanical alloyed W-Cu composite powder," in Proceedings of the 14th International Place Seminar 97, pp. 16–26, Tirol, Austria, May 1997.
62. Y. S. Zu, Y. H. Chiou, and S. T. Lin, "Performance of powder-injection-molded W-4.9Ni-2.1Fe components," Journal of Materials Engineering and Performance, vol. 5, no. 5, pp. 609–614, 1996. · ·
63. Y. S. Zu and S. T. Lin, "Optimizing the mechanical properties of injection molded W-4.9%Ni-2.1% Fe in debinding," Journal of Materials Processing Technology, vol. 71, pp. 337–342, 1997.
64. I. U. Mohsin, C. Gierl, H. Danninger, and M. Momeni, "Thermal de-binding kinetics of injection molded W-8%Ni-2%Cu," International Journal of Refractory Metals and Hard Materials, vol. 29, no. 6, pp. 729–732, 2011. · ·
65. P. Suri, R. M. German, and J. P. de Souza, "Influence of mixing and effect of agglomerates on the green and sintered properties of 97W-2.1Ni-0.9Fe heavy alloys," International Journal of Refractory Metals and Hard Materials, vol. 27, no. 4, pp. 683–687, 2009. · ·
66. J. L. Fan, B. Y. Huang, and X. H. Qu, "Rheologic behavior and sintering characteristic of nanocrystal W-Ni-Fe powder," Journal of Central South University of Technology, vol. 32, no. 1, pp. 66–69, 2001.
67. J. Fan, B. Huang, and X. Qu, "Distortion prediction and control of injection molded tungsten heavy alloys," Journal of Advanced Materials, vol. 36, no. 1, pp. 72–74, 2004.
68. B. Huang, J. Fan, S. Liang, and X. Qu, "The rheological and sintering behavior of W-Ni-Fe nano-structured crystalline powder," Journal of Materials Processing Technology, vol. 137, no. 1–3, pp. 177–182, 2003. · ·
69. Y. M. Li, X. H. Qu, and B. Y. Huang, "Injection molded tungsten heavy alloy," Transactions of Nonferrous Metals Society of China, vol. 8, pp. 576–581, 1998.
70. A. Bose, G. Jerman, and R. M. German, "Rhenium alloying of tungsten heavy alloys," Powder Metallurgy International, vol. 21, no. 3, pp. 9–13, 1989.

71. Y. H. Chiou, Y. S. Zu, and S. T. Lin, "Partition of tungsten in the matrix phase for liquid phase sintered 93%W-4.9%Ni-2.1%Fe," Scripta Materialia, vol. 34, no. 1, pp. 135–140, 1996. · ·
72. R. M. German, "Microstructure of the gravitationally settled region in a liquid-phase sintered dilute tungsten heavy alloy," Metallurgical and Materials Transactions A, vol. 26, no. 2, pp. 279–288, 1995. · ·
73. J. S. Benjamin, "Dispersion strengthened superalloys by mechanical alloying," Metallurgical Transactions, vol. 1, no. 10, pp. 2943–2951, 1970. · ·
74. C. Suryanarayana, "Mechanical alloying and milling," Progress in Materials Science, vol. 46, no. 1-2, pp. 1–184, 2001. · ·
75. B. S. Murty and S. Ranganathan, "Novel materials synthesis by mechanicalalloying/milling," International Materials Reviews, vol. 43, no. 3, pp. 101–106, 1998.
76. H. J. Ryu, S. H. Hong, and W. H. Baek, "Microstructure and mechanical properties of mechanically alloyed and solid-state sintered tungsten heavy alloys," Materials Science and Engineering A, vol. 291, no. 1-2, pp. 91–96, 2000. · ·
77. K. H. Lee, S. I. Cha, H. J. Ryu, and S. H. Hong, "Effect of two-stage sintering process on microstructure and mechanical properties of ODS tungsten heavy alloy," Materials Science and Engineering A, vol. 458, no. 1-2, pp. 323–329, 2007. · ·
78. S. H. Hong and H. J. Ryu, "Combination of mechanical alloying and two-stage sintering of a 93W-5.6Ni-1.4Fe tungsten heavy alloy," Materials Science and Engineering A, vol. 344, no. 1-2, pp. 253–260, 2003. · ·
79. K. H. Lee, S. I. Cha, H. J. Ryu, and S. H. Hong, "ffect of sintering process on microstructure and mechanical properties of ODS tungsten heavy alloy," Materials Science and Engineering A, vol. 452-453, pp. 55–60, 2007.
80. J.-W. Yan, Y. Liu, A.-F. Peng, and Q.-G. Lu, "Fabrication of nano-crystalline W-Ni-Fe pre-alloyed powders by mechanical alloying technique," Transactions of Nonferrous Metals Society of China (English Edition), vol. 19, no. 3, pp. s711–s717, 2009. · ·
81. D. P. Xiang, L. Ding, Y. Y. Li, J. B. Li, X. Q. Li, and C. Li, "Microstructure and mechanical properties of fine-grained

W-7Ni-3Fe heavy alloy by spark plasma sintering," Materials Science and Engineering A, vol. 551, pp. 95–99, 2012. ··

82. Z. W. Zhang, J. E. Zhou, S. Q. Xi, G. Ran, and P. L. Li, "Amorphization and thermal stability of mechanical alloyed W–Ni–Fe," Materials Science and Engineering A, vol. A410, pp. 34–39, 2006.

83. F. Jing-Lian, L. Tao, C. Hui-Chao, and W. Deng-Long, "Preparation of fine grain tungsten heavy alloy with high properties by mechanical alloying and yttrium oxide addition," Journal of Materials Processing Technology, vol. 208, no. 1–3, pp. 463–469, 2008. ··

84. J. Fan, B. Huang, X. Qu, and K. A. Khalil, "MIM of mechanically alloyed nanoscale W-Ni-Fe powder," International Journal of Powder Metallurgy, vol. 38, no. 6, pp. 56–61, 2002.

85. J. S. C. Jang, J. C. Fwu, L. J. Chang, G. J. Chen, and C. T. Hsu, "Study on the solid-phase sintering of the nano-structured heavy tungsten alloy powder," Journal of Alloys and Compounds, vol. 434-435, pp. 367–370, 2007. ··

86. E. P. Kim, M. H. Hong, W. H. Baek, and I. H. Moon, "The effect of manganese addition on the microstructure of W-Ni-Fe heavy alloy," Metallurgical and Materials Transactions A, vol. 30, pp. 627–632, 1999.

87. P. Suri, S. V. Atre, R. M. German, and J. P. de Souza, "Effect of mixing on the rheology and particle characteristics of tungsten-based powder injection molding feedstock," Materials Science and Engineering A, vol. 356, no. 1-2, pp. 337–344, 2003. ··

88. J. L. Fan, X. Gong, B. Y. Huang, M. Song, T. Liu, and J. M. Tian, "Densification behavior of nanocrystalline W-Ni-Fe composite powders prepared by sol-spray drying and hydrogen reduction process," Journal of Alloys and Compounds, vol. 489, no. 1, pp. 188–194, 2010. ··

89. B. Katavi and Z. Odanovi, "Effects of strain aging on the structure and mechanical properties of PM 92·5W-5Ni-2·5Fe heavy alloys," Powder Metallurgy, vol. 48, no. 3, pp. 288–294, 2005. ·

90. U. Ravi Kiran, A. Sambasiva Rao, M. Sankaranarayana, and T. K. Nandy, "Swaging and heat treatment studies on sintered 90W-6Ni-2Fe-2Co tungsten heavy alloy," International Journal

of Refractory Metals and Hard Materials, vol. 33, pp. 113–121, 2012. · ·

91. G. R. Goren-Muginstein and A. Rosen, "The effect of cold deformation on grain refinement of heavy metals," Materials Science and Engineering A, vol. 238, no. 2, pp. 351–356, 1997. · ·

92. Y. Yang, H. Lianxi, and W. Erde, "Microstructure and mechanical properties of a hot-hydrostatically extruded 93W-4.9Ni-2.1Fe alloy," Materials Science and Engineering A, vol. 435-436, pp. 620–624, 2006. · ·

93. Z. Xiaoqing, L. Shukui, L. Jinxu, W. Yingchun, and W. Xing, "Self-sharpening behavior during ballistic impact of the tungsten heavy alloy rod penetrators processed by hot-hydrostatic extrusion and hot torsion," Materials Science and Engineering A, vol. 527, no. 18-19, pp. 4881–4886, 2010. · ·

94. G. Bao and K. T. Ramesh, "Plastic flow of a tungsten-based composite under quasi-static compression," Acta Metallurgica Et Materialia, vol. 41, no. 9, pp. 2711–2719, 1993. · ·

95. Z. Xie, J. Yang, X. Huang, and Y. Huang, "Microwave processing and properties of ceramics with different dielectric loss," Journal of the European Ceramic Society, vol. 19, no. 3, pp. 381–387, 1999. · ·

96. M. Oghbaei and O. Mirzaee, "Microwave versus conventional sintering: a review of fundamentals, advantages and applications," Journal of Alloys and Compounds, vol. 494, no. 1-2, pp. 175–189, 2010. · ·

97. A. Upadhyaya, S. K. Tiwari, and P. Mishra, "Microwave sintering of W-Ni-Fe alloy," Scripta Materialia, vol. 56, no. 1, pp. 5–8, 2007. · ·

98. C. Zhou, J. Yi, S. Luo, Y. Peng, L. Li, and G. Chen, "Effect of heating rate on the microwave sintered W-Ni-Fe heavy alloys," Journal of Alloys and Compounds, vol. 482, no. 1-2, pp. L6–L8, 2009. · ·

99. A. Mondal, A. Upadhyaya, and D. Agrawal, "Microwave and conventional sintering of 90W-7Ni-3Cu alloys with premixed and prealloyed binder phase," Materials Science and Engineering A, vol. 527, no. 26, pp. 6870–6878, 2010. · ·

100. M. Yan and J. Hu, "Microwave sintering of high-permeability $(Ni_{0.20}Zn_{0.60}Cu_{0.20})Fe_{1.98}O_4$ ferrite at low sintering temperatures," Journal of Magnetism and Magnetic Materials, vol. 305, no. 1, pp. 171–176, 2006. · ·
101. Z. A. Munir, U. Anselmi-Tamburini, and M. Ohyanagi, "The effect of electric field and pressure on the synthesis and consolidation of materials: a review of the spark plasma sintering method," Journal of Materials Science, vol. 41, no. 3, pp. 763–777, 2006. · ·
102. R. Orru, R. Licheri, A. M. Locci, A. Cincotti, and G. Cao, "Consolidation/synthesis of materials by electric current activated/assisted sintering," Materials Science and Engineering R: Reports, vol. 63, no. 4–6, pp. 127–287, 2009.
103. Z.-W. Zhang, J.-E. Zhou, S.-Q. Xi, G. Ran, and P.-L. Li, "Phase transformation and thermal stability of mechanically alloyed W-Ni-Fe composite materials," Materials Science and Engineering A, vol. 379, no. 1-2, pp. 148–153, 2004. · ·
104. V. N. Chuvil›deev, A. V. Nokhrin, G. V. Baranov et al., "Study of the structure and mechanical properties of nano- and ultradispersed mechanically activated heavy tungsten alloys," Nanotechnologies in Russia, vol. 8, no. 1-2, pp. 108–121, 2013. ·
105. Y. Li, K. Hu, X. Li, X. Ai, and S. Qu, "Fine-grained 93W-5.6Ni-1.4Fe heavy alloys with enhanced performance prepared by spark plasma sintering," Materials Science and Engineering A, vol. 573, pp. 245–252, 2013. · ·
106. B. H. Rabin and R. M. German, "Microstructure effects on tensile properties of tungsten-Nickel-Iron composites," Metallurgical Transactions A, vol. 19, no. 6, pp. 1523–1532, 1988. · ·
107. J. D. Clayton, "Continuum multiscale modeling of finite deformation plasticity and anisotropic damage in polycrystals," Theoretical and Applied Fracture Mechanics, vol. 45, no. 3, pp. 163–185, 2006. · ·
108. X. Q. Li, H. W. Xin, K. Hu, and Y. Y. Li, "Microstructure and properties of ultra-fine tungsten heavy alloys prepared by mechanical alloying and electric current activated sintering," Transactions of Nonferrous Metals Society of China, vol. 20, no. 3, pp. 443–449, 2010. · ·

109. K. Hu, X. Li, D. Zheng, and Y. Li, "SPS densification behavior of W-5.6Ni-1.4Fe heavy alloy powders," Rare Metals, vol. 30, no. 1, pp. 581–587, 2011. · ·

110. K. Hu, X. Li, S. Qu, and Y. Li, "Spark-Plasma Sintering of W-5.6Ni-1.4Fe Heavy Alloys: Densification and Grain Growth," Metallurgical and Materials Transactions A: Physical Metallurgy and Materials Science, vol. 44, no. 2, pp. 923–933, 2013. · ·

111. K. Vanmeensel, A. Laptev, J. Hennicke, J. Vleugels, and O. van der Biest, "Modelling of the temperature distribution during field assisted sintering," Acta Materialia, vol. 53, no. 16, pp. 4379–4388, 2005. · ·

112. D. Salamon, M. Eriksson, M. Nygren, and Z. Shen, "Can the use of pulsed direct current induce oscillation in the applied pressure during spark plasma sintering?" Science and Technology of Advanced Materials, vol. 13, no. 1, Article ID 015005, 2012. · ·

113. Y. Liu and R. M. German, "Contact angle and solid-liquid-vapor equilibrium," Acta Materialia, vol. 44, no. 4, pp. 1657–1663, 1996. · ·

114. J. Zhu, S. Cao, and H. Liu, "Fabrication of W-Ni-Fe alloys with gradient structures," International Journal of Refractory Metals and Hard Materials, vol. 36, pp. 72–75, 2013. · ·

115. R. Gero, L. Borukhin, and I. Pikus, "Some structural effects of plastic deformation on tungsten heavy metal alloys," Materials Science and Engineering A, vol. 302, no. 1, pp. 162–167, 2001. · ·

116. R. M. German, Liquid Phase Sintering, Plenum Press, New York, NY, USA, 1985.

117. R. Bollina and R. M. German, "Heating rate effects on microstructural properties of liquid phase sintered tungsten heavy alloys," International Journal of Refractory Metals and Hard Materials, vol. 22, no. 2-3, pp. 117–122, 2004. · ·

118. D. V. Edmonds, "Structure/property relationships in sintered heavy alloys," International Journal of Refractory Metals and Hard Materials, vol. 10, no. 1, pp. 15–26, 1991. · ·

119. H. Liu, S. Cao, J. Zhu, Y. Jin, and B. Chen, "Densification, microstructure and mechanical properties of 90W-4Ni-6Mn heavy alloy," International Journal of Refractory Metals and Hard Materials, vol. 37, pp. 121–126, 2013. · ·

120. F. Wakai, M. Yoshida, Y. Shinoda, and T. Akatsu, "Coarsening and grain growth in sintering of two particles of different sizes," Acta Materialia, vol. 53, no. 5, pp. 1361–1371, 2005. · ·
121. T.-K. Kang and D. N. Yoon, "Coarsening of tungsten grains in liquid nickel-tungsten matrix," Metallurgical Transactions A, vol. 9, no. 3, pp. 433–438, 1978. · ·
122. U. R. Kiran, A. Panchal, M. Sankaranarayana, and T. K. Nandy, "Tensile and impact behavior of swaged tungsten heavy alloys processed by liquid phase sintering," International Journal of Refractory Metals and Hard Materials, vol. 37, pp. 1–11, 2013. · ·
123. O. J. Kwon and D. N. Yoon, "Closure of isolated pores in liquid phase sintering of W-Ni,"The International Journal of Powder Metallurgy & Powder Technology, vol. 17, no. 2, pp. 127–131, 1981.
124. D. J. Lee and R. M. German, "Sintering behavior of iron-aluminium powder mixes,"International Journal of Powder Metallurgy and Powder Technology, vol. 21, pp. 9–14, 1985.
125. S.-M. Lee and S.-J. L. Kang, "Theoretical analysis of liquid-phase sintering: pore filling theory," Acta Materialia, vol. 46, no. 9, pp. 3191–3202, 1998. · ·
126. J. Langer, M. J. Hoffmann, and O. Guillon, "Electric field-assisted sintering and hot pressing of semiconductive zinc oxide: a comparative study," Journal of the American Ceramic Society, vol. 94, no. 8, pp. 2344–2353, 2011. · ·
127. D. Demirskyi, H. Borodianska, D. Agrawal, A. Ragulya, Y. Sakka, and O. Vasylkiv, "Peculiarities of the neck growth process during initial stage of spark-plasma, microwave and conventional sintering of WC spheres," Journal of Alloys and Compounds, vol. 523, pp. 1–10, 2012. · ·
128. K. Wang, Z. Fu, W. Wang, Y. Wang, J. Zhang, and Q. Zhang, "Study on fabrication and mechanism in of porous metals by spark plasma sintering," Journal of Materials Science, vol. 42, pp. 302–307, 2007.
129. J. Lankford, H. Couque, A. Bose, and R. German, "Dynamic deformation and failure of tungsten heavy alloys," in Proceedings of the TMS Annual Meeting on Recent Advances in Tungsten and

Tungsten Alloys, pp. 151–160, Warrendale, Pa, USA, February 1991.

130. Y. ahin, Introduction to Composite Materials, Seçkin Publication, Ankara, Turkey, 2nd edition, 2006.

131. F. Akhtar, "An investigation on the solid state sintering of mechanically alloyed nano-structured 90W-Ni-Fe tungsten heavy alloy," International Journal of Refractory Metals and Hard Materials, vol. 26, no. 3, pp. 145–151, 2008. · ·

132. I. S. Humail, F. Akhtar, S. J. Askari, M. Tufail, and X. Qu, "Tensile behavior change depending on the varying tungsten content of W-Ni-Fe alloys," International Journal of Refractory Metals and Hard Materials, vol. 25, no. 5-6, pp. 380–385, 2007. · ·

133. X. Gong, J. L. Fan, B. Y. Huang, and J. M. Tian, "Microstructure characteristics and a deformation mechanism of fine-grained tungsten heavy alloy under high strain rate compression," Materials Science and Engineering A, vol. 527, no. 29-30, pp. 7565–7570, 2010. · ·

134. Y. Yu, L. Hu, and E. Wang, "Microstructure and mechanical properties of a hot-hydrostatically extruded 93W-4.9Ni-2.1Fe alloy," Materials Science and Engineering A, vol. 435-436, pp. 620–624, 2006. · ·

135. A. Upadhyaya, "Processing strategy for consolidating tungsten heavy alloys for ordnance applications," Materials Chemistry and Physics, vol. 67, no. 1–3, pp. 101–110, 2001. · ·

136. L. S. Magness Jr., "High strain rate deformation behaviors of kinetic energy penetrator materials during ballistic impact," Mechanics of Materials, vol. 17, no. 2-3, pp. 147–154, 1994. · ·

137. K. T. Ramesh and R. S. Coates, "Microstructural influences on the dynamic response of tungsten heavy alloys," Metallurgical Transactions A, vol. 23, no. 9, pp. 2625–2630, 1992. ·

138. S. Bless and R. Chau, "Tensile failure of tungsten rods," in Shock Compression of Condensed Matter, M. D. Furnish, M. Elert, T. P. Russell, and C. T. White, Eds., pp. 603–606, American Institute of Physics, New York, NY, USA, 2006.

139. D. P. Dandekar and W. J. Weisgerber, "Shock response of a heavy tungsten alloy,"International Journal of Plasticity, vol. 15, no. 12, pp. 1291–1309, 1999. · ·

140. J. C. F. Millett, N. K. Bourne, Z. Rosenberg, and J. E. Field, "Shear strength measurements in a tungsten alloy during shock loading," Journal of Applied Physics, vol. 86, no. 12, pp. 6707–6709, 1999. · ·

141. M. Zhou and R. J. Clifton, "Dynamic constitutive and failure behavior of a two-phase tungsten composite," Journal of Applied Mechanics, Transactions ASME, vol. 64, no. 3, pp. 487–494, 1997. · ·

142. D.-K. Kim, S. Lee, and H.-S. Song, "Effect of tungsten particle shape on dynamic deformation and fracture behavior of tungsten heavy alloys," Metallurgical and Materials Transactions A, vol. 29, no. 3, pp. 1057–1069, 1998. · ·

143. Q. Wei, T. Jiao, K. T. Ramesh et al., "Mechanical behavior and dynamic failure of high-strength ultrafine grained tungsten under uniaxial compression," Acta Materialia, vol. 54, no. 1, pp. 77–87, 2006. · ·

144. W.-S. Lee, G.-L. Xiea, and C.-F. Lin, "The strain rate and temperature dependence of the dynamic impact response of tungsten composite," Materials Science and Engineering A, vol. 257, no. 2, pp. 256–267, 1998. · ·

145. Y. Zhou, Y. Wang, P. K. Mallick, and Y. Xia, "Strain softening constitutive equation for tungsten heavy alloy under tensile impact," Materials Letters, vol. 58, no. 22-23, pp. 2725–2729, 2004. · ·

146. M. Zhou, R. J. Clifton, and A. Needleman, "Shear band formation in a W-Ni-Fe heavy alloy under plate impact," in Tungsten & Tungsten Alloys, pp. 343–356, Metal Powder Industries Federation, Arlington, Va, USA, 1992.

147. H. Couque Jr., J. Lankford, and A. Bose, "Tensile fracture and shear location under high loading rate in tungsten alloys," Journal de Physique III, vol. 2, pp. 2225–2229, 1992.

148. S. Yadav and T. Ramesh, "Strain rate properties of tungsten based composites," Materials Science and Engineering A, vol. 203, pp. 9140–9153, 1995.

149. R. D. Dick, V. Ramachandran, J. D. Williams, R. W. Armstrong, W. H. Holt, and W. Mock Jr., "Dynamic deformation of W_7Ni_3Fe alloy via reverse-ballistic impact," in Tungsten and Tungsten

Alloys-Recent Advances, A. Crowson and E. S. Chen, Eds., pp. 269–276, The Minerals, Metals and Materails Society, 1991.

150. J. B. Stevens and R. C. Batra, "Adiabatic shear bands in the taylor impact test for a WHA rod," International Journal of Plasticity, vol. 14, no. 9, pp. 841–854, 1998. · ·

151. M. Zhou, "The growth of shear bands in composite microstructures," International Journal of Plasticity, vol. 14, no. 8, pp. 733–754, 1998. · ·

152. R. G. Whirley, B. E. Engelmann, and J. O. Hallquist, "DYNA2D, a nonlinear, explicit, two-dimensional finite element code for solid mechanics, user manual," Lawrence Livermore National Laboratory Report UCRL-MA-110630, 1992.

153. A. M. Rajendran, "High strain rate behavior of metals, ceramics, and concrete," Tech. Rep. WL-TR-92-4006, Wright Patterson Air Force Base, Wright-Patterson Air Force Base, Ohio, USA, 1992.

154. J. B. Stevens, Finite element analysis of adiabatic shearbands in impact and penetration problems [M.S. thesis], Virginia Polytechnic Institute and State University, Blacksburg, Va, USA, 1996.

155. J. D. Clayton, "Dynamic plasticity and fracture in high density polycrystals: constitutive modeling and numerical simulation," Journal of the Mechanics and Physics of Solids, vol. 53, no. 2, pp. 261–301, 2005. · ·

156. J. D. Clayton, "Continuum multiscale modeling of finite deformation plasticity and anisotropic damage in polycrystals," Theoretical and Applied Fracture Mechanics, vol. 45, no. 3, pp. 163–185, 2006. · ·

157. J. D. Clayton, "Modeling dynamic plasticity and spall fracture in high density polycrystalline alloys," International Journal of Solids and Structures, vol. 42, no. 16-17, pp. 4613–4640, 2005. · ·

158. J. D. Clayton, "Plasticity and spall in high density polycrystals: modeling and simulation," inCompression of Condensed Matter, M. D. Furnish, M. Elert, T. P. Russell, and C. T. White, Eds., pp. 311–314, American Institute of Physics, New York, NY, USA, 2006.

159. T. J. Vogler and J. D. Clayton, "Heterogeneous deformation and spall of an extruded tungsten alloy: plate impact experiments and crystal plasticity modeling," Journal of the Mechanics and Physics of Solids, vol. 56, no. 2, pp. 297–335, 2008.
160. T. H. Antoun, L. Seaman, D. R. Curran, G. I. Kanel, S. V. Razorenov, and A. V. Utkin, Spall Fracture, Springer, New York, NY, USA, 2003.
161. D. P. Dandekar and W. J. Weisgerber, "Shock response of a heavy tungsten alloy," International journal of plasticity, vol. 15, no. 12, pp. 1291–1309, 1999. ··
162. D. P. Dandekar, "Spall strength of silicon carbide under normal and simultaneous compression-shear shock wave loading," International Journal of Applied Ceramic Technology, vol. 1, no. 2, pp. 261–268, 2004.
163. Y. I. Mescheryakov, "Meso-macro energy exchange in shock deformed and fractured solids," in High-Pressure Shock Compression of Solids VI: Old Paradigms and New Challenges, Y. Horie, L. Davison, and N. N. Thadhani, Eds., pp. 169–213, Springer, New York, NY, USA, 2003.

Chapter 6

Synthesis of β-SiC Fine Fibers by the Forcespinning Method with Microwave Irradiation

Alfonso Salinas[1], Maricela Lizcano[2], and Karen Lozano[1]

[1]Mechanical Engineering Department, The University of Texas-Pan American, Edinburg, TX 78539, USA

[2]National Aeronautics and Space Administration, Materials and Structures Division, Glenn Research Center at Lewis Field, Cleveland, OH 44135, USA

ABSTRACT

A rapid method for synthesizing β-silicon carbide (β-SiC) fine fiber composite has been achieved by combining forcespinning technology with microwave energy processing. β-SiC has applications as composite reinforcements, refractory filtration systems, and other high temperature applications given their properties such as low density, oxidation resistance, thermal stability, and wear resistance. Nonwoven fine fiber mats were prepared through a solution based method using polystyrene (PS) and polycarbomethylsilane (PCmS) as the precursor materials. The fiber spinning was performed under different parameters to obtain high yield, fiber homogeneity, and small diameters. The spinning was carried out under controlled nitrogen environment to control and reduce oxygen content. Characterization was conducted using scanning electron microscopy (SEM), X-ray diffraction (XRD), and Fourier transform infrared spectroscopy (FTIR). The results show high yield, long continuous bead-free nonwoven fine fibers with diameters ranging from 270 nm to 2 μm depending on the selected processing parameters. The fine fiber mats show formation of highly crystalline β-SiC fine fiber after microwave irradiation.

INTRODUCTION

In the past 50 years, microwave irradiation has been utilized to process various materials such as semiconductors and inorganic and polymeric materials. More recently, microwave energy has been used to sinter powdered metals as well as ceramic systems [1]. Microwaves are electromagnetic radiation with a wavelength from 1 mm to 1 m with frequencies in the range of 1 to 300 GHz [2]. Most common microwaves furnaces used for industrial and scientific applications operate at a frequency of 2.45 GHz [3]. The most effective way to produce microwaves is from a magnetron source, but they can also be produced from klystrons, power grid tubes, traveling wave tubes, and gyrotrons [4].

Microwave processing of ceramics was first reported in 1968 by Tinga and Voss [5]. However, it was not until the 1980s that high temperature processing with microwave energy started gaining much ground [6]. Although microwave processing of advanced ceramic materials is still developing, it offers many advantages over conventional ceramic processing methods such as reduced heating times and lower power consumption [7].

In traditional thermal material processing, energy is transferred by convection and radiant heat onto the surface of the material and then through the material by conduction heating. Materials requiring long processing times via traditional methods undergo thermal gradients within the material, wherein the surface of the material is exposed to more heat than the core of the material, resulting in surface damage [8]. However, in microwave thermal processing, heat is directly transferred to the material volumetrically by molecular interactions with the electromagnetic field. Since the diffusion of heat through the surface, as in traditional thermal processing, is bypassed by volumetric heating, uniform heating and fast processing times can be achieved with heating rates as high as 1000°C/min [8–10].

The major advantages of microwave processing over conventional heating methods are reduced thermal gradients within the material, faster reaction times, lower processing temperatures, high density microstructures, and improved mechanical properties [11, 12]. These advantages strongly support the use of microwave thermal processing for advanced materials development.

The forcespinning (FS) process is a rapid method to produce nanosize to micron size fibrous materials. Unlike electrospinning, yields as high as 1 g/min in a lab scale unit are easily achieved. The combination of this fiber making technology with the fast heating rates of MW irradiation provides a rapid route for producing ceramic materials. FS utilizes centrifugal forces to overcome shear forces promoting fiber elongation. Process, theory, and schematics have been reported elsewhere [13–15]. In this research, the development and optimization of the parameters involved in the production of β-SiC fine fibers were carefully analyzed and developed materials characterized. The prepared green fine fibers were spun from polymeric precursors. β-SiC nanomaterials have been intensively studied given their unique properties such as high mechanical strength, high thermal conductivity, low thermal expansion coefficient, and chemical inertness when compared to those of their bulk counterparts [16]. In this case, the utilization of a fibrous morphology also provides a significant increase in surface to volume ratio. Many studies have shown the potential applications of β-SiC fine fibers and the lab scale results have shown promising applications; therefore, scientists are researching new and easier methods to develop β-SiC nanostructures.

The preceramic fine fibers were developed utilizing a solution of polystyrene and polycarbomethylsilane. The developed nonwoven fine fiber mats were characterized by FESEM (field emission scanning electron microscope), XRD (X-ray diffraction), EDS (energy dispersive spectroscopy), and FTIR (Fourier transform infrared spectroscopy).

EXPERIMENTAL

Materials

Polystyrene (PS) with a molecular weight of 280,000 g/mol and polycarbomethylsilane (PCmS) with a molecular weight of 800 g/mol were purchased from Sigma-Aldrich (Milwaukee, WI, USA) and used as received. Toluene was purchased from Fisher Scientific (Waltham, MA, USA) and used as received. The PS/PCmS/Toluene (15, 20, and 25 wt% of PS with a 2:1 ratio of PS:PCmS) solutions were prepared inside a MBRAUN (Stratham, NH) glovebox under nitrogen atmosphere in order to prevent oxidation. The solutions were prepared in 20 mL scintillation vials and sealed with parafilm to prevent solvent evaporation. Solutions were magnetically stirred for a period of 4 hours.

Fine Fiber Development

A FS system was placed inside a glovebox under nitrogen environment. Approximately 2 mL of solution was inserted into a cylindrical type spinneret using a 10 mL syringe. The spinneret was outfitted with 30 gage needles. The angular velocity at which the fibers were spun was varied from 5,000 rpm to 9,000 rpm. The solution was depleted in less than 30 sec. The fibers were deposited on a circular collector having 16 equally spaced polytetrafluoroethylene (PTFE) bars. The developed fibers were stored in a glovebox to prevent fiber oxidation.

Fibers were cross-linked to maintain geometric integrity of the precursor fine fibers because PS reaches the glass transition temperature before the preceramic polymer (PCmS) is converted to ceramic SiC during heat treatment. The collected fine fibers were placed under a 254 nm wavelength UV light source for a period of 24 hours. The cross-linking was performed in a glovebox under nitrogen environment.

Microwave Pyrolysis

A microwave furnace, Hi-Tech single mode microwave applicator, was used. The system is fitted with a MH 2.0 W-S water cooled magnetron head assembly. It supplies 2 kW of adjustable microwave energy at

2.45 GHz. The magnetron outputs into a WR340 waveguide. An Omega iSeries iR2 infrared pyrometer was used to record the temperature. The samples were placed between small SiC susceptor plates which absorb electromagnetic energy and convert it to heat. The sample chamber was fitted with a quartz tube attached to a turbo pump. The microwave heating was carried out under nitrogen gas after evacuation of air with the turbo pump. Power was increased 100 W every 4 minutes up to 600 W. The total processing time at 600 W was 3 minutes. The temperature was observed to be approximately 1140°C at 600 W.

Fiber Characterization

Fiber morphology was analyzed using the Carl Zeiss Sigma VP scanning electron microscope. Fiber diameters were measured using the Carl Zeiss AxioVision software. For the X-ray diffraction analysis, a Bruker AXS D8 diffractometer was utilized. The fine fibers were scanned from 20 to 80° (2θ angles) using a 2D detector. FTIR-ATR with a diamond tip was carried out using an Agilent Technologies Cary 600 Series FTIR spectrometer.

RESULTS AND DISCUSSION

The optimization of the SiC fiber precursors was conducted in a previous study [17]. This study focused on developing highly crystalline β-SiC fine fibers through conventional pyrolysis methods. Several parameters were evaluated that resulted in fibers with average diameters ranging from 270 nm to 2 μm. It was concluded that the parameters that synergistically contributed to the development of homogeneous, high yield, bead-free continuous green SiC fibers were of a polymer concentration of 20 wt% in the developed solution which was then processed at an angular velocity of 7000 rpm. Figure 1 shows a nonwoven fine fiber mat with its observable corresponding fiber diameter distribution. Figure 2 shows an SEM micrograph of the precursor fibers showing micron and submicron size fibers. Fiber-fiber adhesion can be observed as indicated by the red boxes. The above mentioned fiber spinning parameters were selected to prepare samples utilized in this work to further analyze the effect of heat treatment via microwave energy processing.

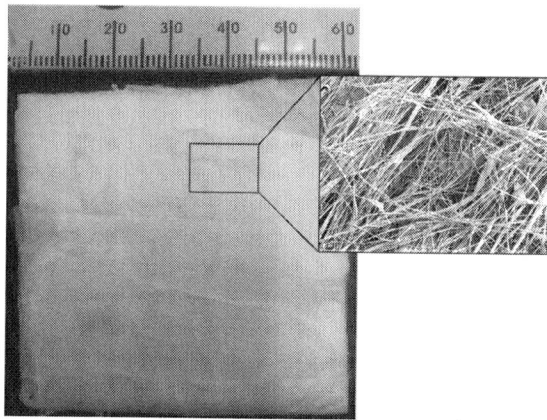

Figure 1: 6 cm × 6 cm SiC precursor mat obtained after 30 sec.

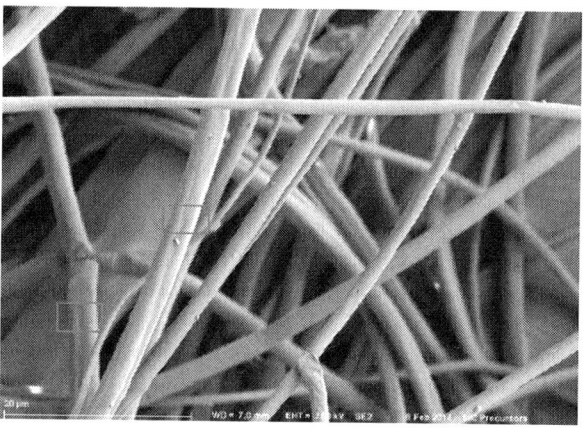

Figure 2: SEM micrograph of SiC precursors. Red boxes show fiber surface roughness.

The materials in this study consisted of PS and PCmS with molecular formulas (C_8H_8) and $(C_2H_6Si)_n$, respectively. At temperatures between 550°C and 800°C, the precursor becomes an inorganic material as it begins to decompose Si–H, Si–CH_3, Si–CH_2–Si, H_2, CH_4, CO, and CO_2 species which are eliminated from the precursor [18–20]. Increasing the temperature from 800°C up to 1000°C results in amorphous SiC while crystalline SiC begins to form at 1000°C with the evolution of H_2 [18].

XRD results are shown in Figure 3 with a corresponding peak list given in Table 1. The peaks indicate conversion of precursor to -SiC with 2θ = 35.62°, 41.42°, 59.95°, and 71.74° (reference code: 00-029-1129) ascribed to lattice planes (111), (200), (220), and (311), respectively [21]. Additionally, the slight broadening base of the peaks is indicative of either unreacted material or some amorphous SiC remaining in the sample. A small carbon peak is noted at 2θ = 26.74° (reference code: 04-014-0337).

Table 1: Corresponding XRD peak list from Figure 3

Pos. [°2Th.]	Height[cts]	d-spacing[Å]	Rel. Int.[%]	Matched by
26.7407	197.39	3.33386	9.92	04-014-0337
35.6177	1989.65	2.5207	100	00-029-1129
41.4258	111.8	2.17973	5.62	00-029-1129
59.9538	501.03	1.54295	25.18	00-029-1129
71.7441	328.15	1.31564	16.49	00-029-1129

Pos = position of the peak on the 2θ axis in the XRD spectra.

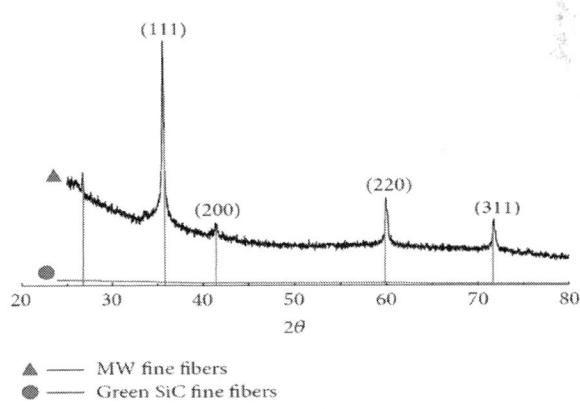

Figure 3: XRD spectra of fine fibers before and after microwave assisted heating.

SiC FTIR reflectance bands can be seen between wave numbers 740 and 970 cm^{-1} [22–24]. The FTIR scan shown in Figure 4 depicts a reflectance band at wave number 788 cm^{-1}, supporting ceramic conversion of precursor fine fibers shown in Figure 3. The peak at wave number 1003 cm^{-1} corresponds to Si–O bonds indicating the presence of SiO_2 in the sample [25, 26]. Although the SiO_2 was not detected in XRD results, it may be due to the nanosized geometries of its content. If the uncured precursor is exposed to oxygen during handling, oxidation of Si–H and Si–CH$_3$ can occur resulting in the formation of Si–O–Si bonds [27]. The available oxygen species in the precursor can then form amorphous SiO_2 at temperatures below 1200°C [19]. EDS was performed in the precursor fiber as well as in the ceramic converted fibers. The small presence of oxygen in both samples is given in Table 2. Reflectance band at 2156 cm^{-1} is consistent with MeSiH$_3$ [28] while bands seen at 2048 cm^{-1} and 1963 cm^{-1} indicate the presence of CO species [29].

Table 2: EDS of SiC precursor and microwaved fine fibers

Element	SiC precursor NFs		Microwaved NFs	
	wt%	at%	wt%	at%
Si	17.63	8.71	37.19	21.32
C	68.99	79.69	46.29	62.06
O	13.38	11.6	16.52	16.62

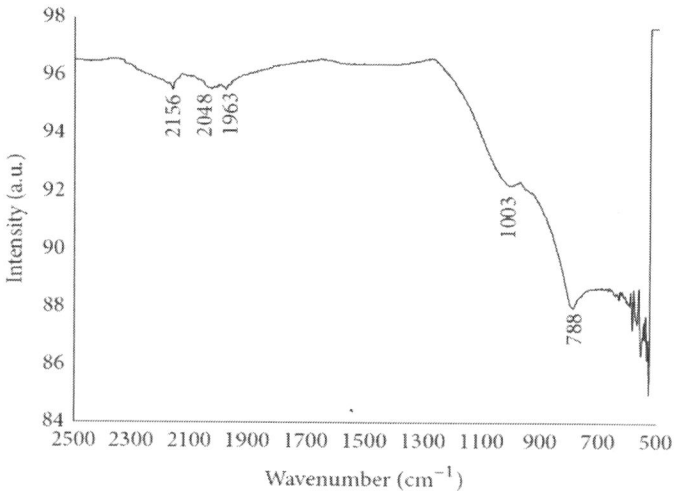

Figure 4: FTIR of SiC fine fibers after microwave irradiation.

These types of composite materials have applications in nanoelectronics, nanomechanics, reinforced composite materials, and nanosensors to mention some [30]. More importantly, these materials may be used to improve SiC/SiO_2 bonding interfaces for improved performance in electronic devices [31, 32]. materials have been previously synthesized including the use of microwave irradiation [21, 33–36]. However, in these studies Si, SiO_2, and graphite powders were used as the precursor materials rather than a preceramic polymer. An SEM micrograph of the converted ceramic fine fiber is shown in Figure 5. Micron and submicron size fibers can be seen in this figure. Also, noted is a rough fiber structure in some areas which may be due to adhesion of two or more fibers forming ribbon-like fibers or shrinkage that a fiber undergoes as a result of volatilization of solvent during microwave processing as previously observed in annealed fibers [37]. The micrograph also shows many short fibers, although fibers are expected to be up to 6 feet in length [14]. The precursor fine fibers are removed with tweezers resulting in broken fibers before being placed on the susceptor for MW processing. A continuous deposition method with subsequent MW treatment will preserve fiber length. Figure 6 shows a SEM micrograph of a single SiC nanofiber.

Figure 5: SEM micrograph of ceramic converted fine fibers.

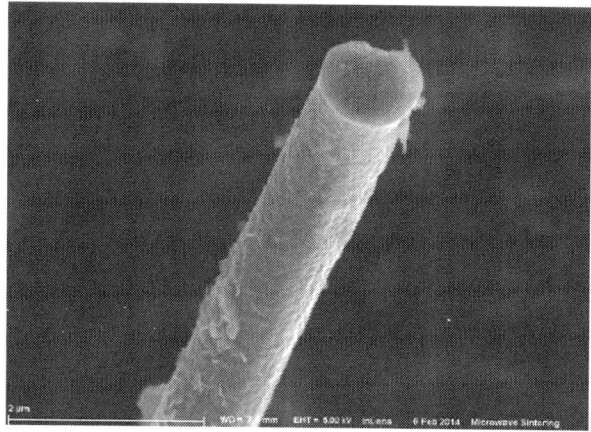

Figure 6: High magnification SEM micrograph of a β-SiC fine fiber.

CONCLUSIONS

In this study, we successfully show ceramic conversion of spun PS/PCmS fine fibers to β-SiC fine fibers utilizing forcespinning technology for rapid fiber development and microwave energy processing for rapid ceramic conversion of spun fibers. The combination of these two technologies illustrates a processing route that can be utilized to produce rapid novel ceramic nanomaterials. XRD and FTIR results confirmed ceramic conversion of the fine fibers. Both micron and submicron fibers were observed in SEM images.

REFERENCES

1. A. Chang, H. Zhang, Q. Zhao, and B. Zhang, Microwave Sintering of Thermistor Ceramics, InTech, 1980.
2. J. D. Katz, "Microwave sintering of ceramics," Annual Review of Materials Science, vol. 22, no. 1, pp. 153–170, 1992.
3. D. K. Agrawal, "Microwave processing of ceramics," Current Opinion in Solid State and Materials Science, vol. 3, no. 5, pp. 480–485, 1998.

4. S. Chandrasekaran, S. Ramanathan, and T. Basak, "Microwave material processing—a review," AIChE Journal, vol. 58, no. 2, pp. 330–363, 2012.
5. W. R. Tinga and W. A. G. Voss, Microwave Power Engineering, edited by E. C. Okress, Academic Press, New York, NY, USA, 1968.
6. D. Agrawal, "Microwave sintering of ceramics, composites and metallic materials, and melting of glasses," Transactions of the Indian Ceramic Society, vol. 65, no. 3, pp. 129–144, 2006.
7. D. E. CLark, D. C. Folz, C. E. Folgar, and M. M. Mahmoud, Microwave Solutions for Ceramic Engineers, The American Ceramic Society, Westerville, Ohio, USA, 2005.
8. E. T. Thostenson and T.-W. Chou, "Microwave processing: fundamentals and applications," Composites Part A: Applied Science and Manufacturing, vol. 30, no. 9, pp. 1055–1071, 1999.
9. Y. C. Kim, C. H. Kim, and D. K. Kim, "Effect of microwave heating on densification and $\alpha \rightarrow \beta$ phase transformation of silicon nitride," Journal of the European Ceramic Society, vol. 17, no. 13, pp. 1625–1630, 1997.
10. S. Somiya, F. Aldinger, R. M. Spriggs, K. Uchino, K. Koumoto, and M. Kaneno, Handbook of Advanced Ceramics: Materials, Applications, Processing, and Properties, Academic Press, 2003, http://books.google.com.eg/books?id=SMCRJi52OcQC&redir_esc=y.
11. D. D. Upadhyaya, A. Ghosh, G. K. Dey, R. Prasad, and A. K. Suri, "Microwave sintering of zirconia ceramics," Journal of Materials Science, vol. 36, no. 19, pp. 4707–4710, 2001.
12. M. Oghbaei and O. Mirzaee, "Microwave versus conventional sintering: a review of fundamentals, advantages and applications," Journal of Alloys and Compounds, vol. 494, no. 1-2, pp. 175–189, 2010.
13. B. Vazquez, H. Vasquez, and K. Lozano, "Preparation and characterization of polyvinylidene fluoride nanofibrous membranes by forcespinning," Polymer Engineering and Science, vol. 52, no. 10, pp. 2260–2265, 2012.
14. S. Padron, A. Fuentes, D. Caruntu, and K. Lozano, "Experimental study of nanofiber production through forcespinning," Journal of Applied Physics, vol. 113, no. 2, Article ID 024318, 2013.

15. K. Sarkar, C. Gomez, S. Zambrano et al., "Electrospinning to forcespinning," Materials Today, vol. 13, no. 11, pp. 12–14, 2010.
16. Z.-M. Huang, Y.-Z. Zhang, M. Kotaki, and S. Ramakrishna, "A review on polymer nanofibers by electrospinning and their applications in nanocomposites," Composites Science and Technology, vol. 63, no. 15, pp. 2223–2253, 2003.
17. A. Salinas, Mass production of β-silicon carbide nanofibers by the novel [M.S. thesis], The University of Texas Pan American, 2013.
18. P. Colombo, R. Riedel, G. D. Soraru, and H.-J. Kleebe, Eds., Polymer Derived Ceramics: From Nano-Structure to Applications, DEStech Publications, 2010.
19. K. Okamura, "Ceramic fibres from polymer precursors," Composites, vol. 18, no. 2, pp. 107–120, 1987.
20. S. Somiya, F. Aldinger, R. M. Spriggs, K. Uchino, K. Koumoto, and M. Kaneno, Eds., Handbook of Advanced Ceramics: Materials, Applications, Processing and Properties, vol. 2, Academic Press, New York, NY, USA, 2013.
21. J. Wang, S. Liu, T. Ding, S. Huang, and C. Qian, "Synthesis, characterization, and photoluminescence properties of bulk-quantity β-SiC/SiO$_x$ coaxial nanowires," Materials Chemistry and Physics, vol. 135, no. 2-3, pp. 1005–1011, 2012.
22. S. B. Qadri, A. W. Fliflet, A. Imam, B. B. Rath, and E. P. Gorzkowski, "Silicon carbide synthesis from agricultural waste," US 20130272947Al, 2013.
23. J. P. Li, A. J. Steckl, I. Golecki et al., "Structural characterization of nanometer SiC films grown on Si," Applied Physics Letters, vol. 62, no. 24, pp. 3135–3137, 1993.
24. M. Perný, V. Šály, M. Váry, and J. Huran, "Electrical and structural properties of amorphous silicon carbide and its application for photovoltaic heterostructures," Elektroenergetika, vol. 4, no. 3, 2011.
25. J. Bullot and M. P. Schmidt, "Physics of amorphous silicon-carbon alloys," Physica Status Solidi (B) Basic Research, vol. 143, no. 2, pp. 345–418, 1987.

26. J. Aguilar, L. Urueta, and Z. Valdez, "Polymeric synthesis of silicon carbide with microwaves," Journal of Microwave Power and Electromagnetic Energy, vol. 40, no. 3, pp. 145–154, 2007.
27. B. Shokri, M. A. Firouzjah, and S. Hosseini, "FTIR analysis of silicon dioxide thin film deposited by metal organic-based PECVD," in Proceedings of the International Plasma Chemistry Society (IPCS ‹09), 32009 ISPC19, Shahid Beheshti University, Bochum, Germany, 2009.
28. W. J. Miller, High temperature oxidation of silicon carbide [M.S. thesis], Air Force Institute of Technology, Wright Paterson Air Force Base, Ohio, USA, 1972.
29. A. L. Smith and N. C. Angelotti, "Correlation of the SiH stretching frequency with molecular structure," Spectrochimica Acta, vol. 15, pp. 412–420, 1959.
30. N. Koizumi, K. Murai, S. Tamayama, H. Kato, T. Ozaki, and M. Yamada, "Diffuse reflectance IR spectroscopic study on the role of promoters in the reactivity of carbon monoxide with hydrogen over novel Pd sulfide catalyst," Fuel Chemistry Division Preprints, vol. 47, no. 2, p. 520, 2002.
31. H. F. Zhang, C. M. Wang, and L. S. Wang, "Helical crystalline SiC/SiO_2 core-shell nanowires," Nano Letters, vol. 2, no. 9, 2002.
32. J. G. Wang, S. Liu, T. Ding, S. Huang, and C. Qian, "Synthesis, characterization, and photoluminescence properties of bulk-quantity of β-SiC/SiO_x coaxail nanowires," Materials Chemistry and Physics, vol. 135, pp. 1005–1011, 2012.
33. D. M. Wolfe, B. J. Hinds, F. Wang et al., "Thermochemical stability of silicon-oxygen-carbon alloy thin films: a model system for chemical and structural relaxation at SiC-SiO_2 interfaces," Journal of Vacuum Science and Technology A: Vacuum, Surfaces and Films, vol. 17, no. 4, pp. 2170–2177, 1999.
34. O.-S. Kwon, S.-H. Hong, and H. Kim, "The improvement in oxidation resistance of carbon by a graded SiC/SiO_2 coating," Journal of the European Ceramic Society, vol. 23, no. 16, pp. 3119–3124, 2003.
35. H. Zhao, Z. Fu, C. Tang, X. Liu, Z. Li, and K. Zhang, "Study of SiC/SiO_2 oxidation-resistant coatings on matrix graphite for HTR fuel element," Nuclear Engineering and Design, vol. 271, pp. 217–220, 2014.

36. Z. He, R. Tu, H. Katsui, and T. Goto, "Synthesis of SiC/SiO$_2$ core-shell powder by rotary chemical vapor deposition and its consolidation by spark plasma sintering," Ceramics International, vol. 39, no. 3, pp. 2605–2610, 2013.

37. Y. Rane, A. Altecor, N. S. Bell, and K. Lozano, "Preparation of superhydrophobic Teflon AF 1600 sub-micron fibers and yarns using the Forcespinning technique," Journal of Engineered Fibers and Fabrics, vol. 8, no. 4, pp. 88–95, 2013.

Chapter 7

Foaming Behaviour, Structure, and Properties of Polypropylene Nanocomposites Foams

M. Antunes, V. Realinho, and J. I. Velasco

Departament de Ciència del Materials i Enginyeria Metal.lúrgica, Centre Català del Plàstic, Universitat Politècnica de Catalunya, C/ Colom 114, 08222 Terrassa, Barcelona, Spain

ABSTRACT

This work presents the preparation and characterization of compression-moulded montmorillonite and carbon nanofibre-polypropylene foams. The influence of these nanofillers on the foaming behaviour was analyzed in terms of the foaming parameters and final cellular structure and morphology of the foams. Both nanofillers induced the formation of a more isometric-like cellular structure in the foams, mainly observed for the MMT-filled nanocomposite foams. Alongside their crystalline characteristics, the nanocomposite foams were also characterized and compared with the unfilled ones regarding their dynamic-mechanical thermal behaviour. The nanocomposite foams showed higher specific storage moduli due to the reinforcement effect of the nanofillers and higher cell density isometric cellular structure. Particularly, the carbon nanofibre foams showed an increasingly higher electrical conductivity with increasing the amount of nanofibres, thus showing promising results as to produce electrically improved lightweight materials for applications such as electrostatic painting.

INTRODUCTION

Although the increasing interest in the preparation and study of polyolefin foams, there is still lack of information regarding the characterization of rigid polypropylene foams thought for structural applications with typical relative densities, that is, the density of the foam divided by that of the respective solid, higher than 0.1. Hence the interest in preparing and studying new polypropylene-based foams by carefully controlling the expansion and final cellular structure [1].

Nowadays, the use of low-density PP foams is rather limited when compared to PE to situations where higher service temperatures or thermal stabilities are required. PE is cheaper and displays a wider range of molecular architectures, making it easier to reach the high melt strengths and extensibilities required for foaming. Also, the rubbery plateau of the polymer melt can be easily increased via cross-linking, that way it wides the optimum temperature window for stable foam production [1]. Contrarily, PP's linear structure makes it harder to foam due to its intrinsically low melt strength [2]. In order to achieve the high expansions required for applications such as packaging, PP is often blended with other polyolefins, mainly low melting point ethylene copolymers such as EVA or ethylene-octene copolymers [3] or used as random copolymer with low ethylene content [4].

Some of the advantages of PP, such as its higher stiffness, strength, and better impact strength, only start to be relevant at higher foam densities ($\rho < 100$ kg/m^3). That is why medium density PP foams have been considered in this work. Nevertheless, even at these relatively high densities, the use of PP requires the improvement of its melt resistance. This was possible with the development of long-chain branching modified grades, conventionally known as high melt strength polypropylenes (HMS-PP). The use of these polypropylenes has been shown to improve the volume expandability and cell uniformity, retard cell coalescence, and increase the expansion ratio, globally broadening the optimum foaming processing window [5, 6].

Four basic foaming processes are commonly used to produce PP foams: (1) direct extrusion, where a foam is directly obtained by sudden decompression at the exit of an extrusion die [7, 8]; (2) injection, where expansion is adjusted inside a closed injection mould; (3) compression moulding, where the material is foamed by simultaneously applying heat and pressure and later expanding the material by sudden decompression, conventionally using exothermic chemical blowing agents such as azodicarbonamide (ADC) [9]; (4) batch foaming, where the material is foamed by initially dissolving N_2 or CO_2 in the solid polymer inside high pressure reactors and afterwards expanding the material by heating at low pressure above the glass transition temperature of the polymer-gas mixture or by sudden pressure drop [10–12].

Due to its versatility, compression moulding was used in this work for the preparation of the medium-density polypropylene foams. It is a process that allows to control the expansion by varying the amount of ADC and processing parameters. Therefore, it enables the analysis of incorporating nanometric-sized reinforcements on the foaming behaviour, cellular structure, and final properties of the material and the effect of the foaming process on the particles' distribution and dispersion [6, 13]. With the disadvantage of presenting solid residues inherent to the thermal decomposition of ADC or an anisotropic cellular distribution with cells smaller close to the surface, foams produced using this technique may reach thicknesses as high as 10 cm with cell sizes in the micrometer range [8]. Comparatively, gas dissolution is a very time-consuming process due to the high times required for dissolving the gas, and the Mucell injection foaming process is rather limited to high density foams (>300 kg/m³) [14]. Albeit the small moulds used in this work, these can be easily scaled-up to produce very complex foamed elements and components by replacing the mould, while in the case of injection, mould replacement would be very costly.

During the last couple of years, polymer nanocomposite foams have received increasing attention in both scientific and industrial communities [15]. It has been proven that small amounts of finely-dispersed nanoparticles may act as sites for bubble nucleation during the foaming process. Particularly, the cell density has been found to increase linearly with the clay concentration for low clay values [16, 17]. Besides, the highest cell density was obtained when the clay platelets were exfoliated, attained to a higher effective particle concentration and thus higher nucleation efficiency [16, 18]. In accordance with the higher cell densities, smaller cell sizes were obtained in the presence of the nanoparticles. Thus, the presence of exfoliated nanoparticles may result in finer cellular structures due to a combined bubble nucleation and melt strain hardening effects [19]. The nanometric size of the particles also increases the interaction with the polymer matrix, offering a high potential for local reinforcement, resulting in macroscopic mechanical enhancements. If one considers the micrometer or submicrometer thickness of the cell walls in foams, the extremely small size of the nanoparticles could locally act reinforcing them. In the case of layered-like nanoparticles such as montmorillonite, good barrier properties can also be expected by the nanosized-platelets limiting gas diffusion during the expansion and stabilization of the foam [15].

Several works have compared the mechanical properties of PVC and PS nanocomposite foams with that of the respective unfilled ones under tensile and compressive conditions [18, 20]. In all cases, using layered silicates such as montmorillonite or carbon nanofibres resulted in higher moduli and tensile strengths; in some cases the nanocomposite foams even display higher specific moduli than the solid unfilled material [21].

Therefore, the specific properties of the foams could be extended with the incorporation of low amounts of functional inorganic phases with high specific surface areas. With that in mind, two types of nanometric-sized reinforcements, montmorillonite, MMT [6, 22] and different amounts of carbon nanofibres, CNF [23, 24], were added to a PP-based foaming formulation, the nanocomposite materials later chemically foamed by compression moulding. The particular case of incorporating conductive nanofillers such as carbon nanofibres could result in the improvement of properties such as the electrical conductivity [25], thus contributing to the development of new lightweight electrically conductive materials.

MATERIALS AND COMPOUNDING

Nanocomposite Preparation

A PP material specifically formulated for foaming applications, referred to as PP, was prepared by meltcompounding using a corotating twin-screw extruder (Collin Kneter 25X36D, L/D = 36), 50 phr of a PP-HMS, and 50 phr of an extrusion grade-type with stearic acid (0.2 phr), talc (1.0 phr), and two different amounts of a chemical blowing agent, azodicarbonamide (1.5 and 3.5 phr). These two concentrations of ADC were used to reach the desired expansion ratios. A constant temperature of 165°C and screw speed of 160 rpm were used for all the materials.

The PP-HMS used was an especially modified long-chain branched PP with a density of 0.902 g·cm^{-3} and melt flow index (MFI) of 2.1 g/10 min at 230°C and 2.16 kg. The linear extrusion-grade type of PP had a density of 0.905 g·cm^{-3}, and MFI of 5.8 g/10 min (230°C and 2.16 kg). The azodicarbonamide (Porofor ADC/M-C1), with an ADC content of 99.1%, a density of 1.65 g·cm^{-3} and an average particle size of 3.9 ± 0.6 µm, was added to the polymer blend in the extruder.

In the case of the montmorillonite nanocomposite, a commercial masterbatch of 75 wt.% of PP with 25 wt.% of an octadecyl amine modified montmorillonite (Nanomer C32P) was melt-compounded with the PP material in the extruder so as to obtain a final nanocomposite with 5.0 phr of the modified montmorillonite (PP-MMT).

Three different carbon nanofibre-polypropylene nanocomposites (PP-CNF) were prepared by melt mixing in the twin screw extruder 5, 10, and 20 wt.% of carbon nanofibres with the previously mentioned PP matrix. The carbon nanofibres used in this work were highly graphitized submicron vapour grown carbon nanofibres, with a typical diameter of 20–80 nm, fibre length >30 µm, density of 1.97 g·cm^{-3}, specific surface area BET (N_2) of $150-200 \text{ m}^2\text{·g}^{-1}$, and electrical resistivity of $10^{-3} \text{ }\Omega\text{·m}$. The reason behind the preparation of these three nanocomposites lies on the interest in studying the electrical conductivity as a function of the amount of nanofibres.

The rather low processing temperatures (165°C) and high screw speeds (160 rpm) prevented the azodicarbonamide from thermally decomposing inside the extruder.

Foaming Process

Prior to foaming by thermal decomposition of the ADC (Figure 1(a)), solid discs of the different materials with a thickness of 3.5 mm and diameter of 74 mm were prepared by compression moulding the extruded pellets in a hot-plate press IQAP-LAP PL-15.

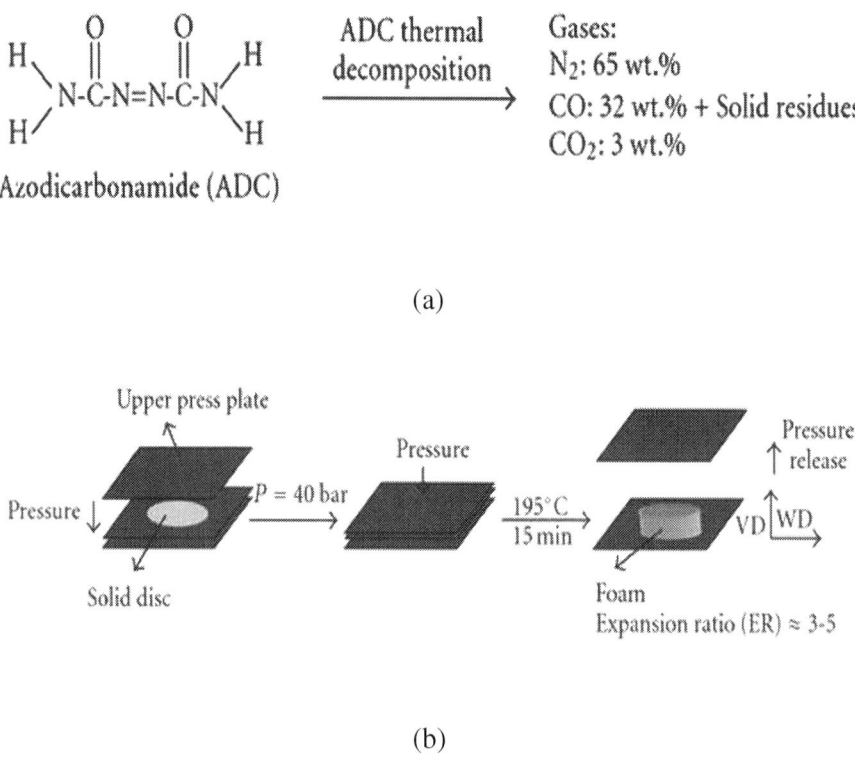

Figure 1: (a) Thermal decomposition of the ADC and (b) one-step compression moulding foaming process schematic. VD: Vertical direction of foaming; WD: Width direction.

A one-step compression moulding process was used to foam the solid discs by placing them inside a circular mould (Φ= 74 mm) and heating at 195C applying a pressure of 40 bar for 15 min using the hot-plate press (see Figure 1(b)).

Testing Procedure

Density of the several solids and foams was measured according to standard procedures (ISO 845).

The cellular structure of the foams was analyzed by scanning electron microscopy, SEM (JEOL JSM-5610). Samples were fractured at low temperature and made conductive by depositing a thin layer of gold. The average cell size (ϕ) and cell density were obtained using the intercept counting method [26]. Two different cell sizes were determined using the procedure presented in [13]: ϕ_{VD} (VD: Vertical Direction), that is, the average cell size in the direction of pressure release and ϕ_{WD} (Width Direction). The aspect ratio, AR (AR = ϕ_{VD}/ϕ_{WD}), was determined using a representative cell population. Schematics showing specimen configurations and most characteristic cellular structure parameters are presented in Figure 2.

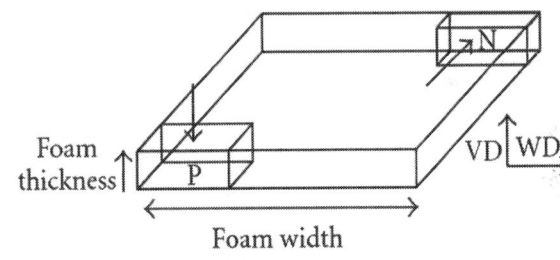

(a)

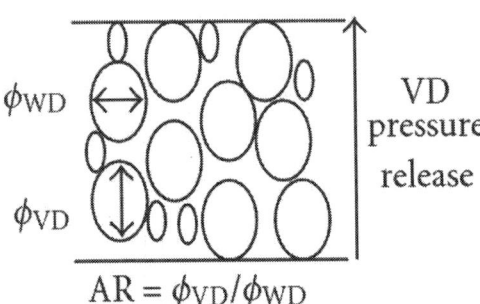

(b)

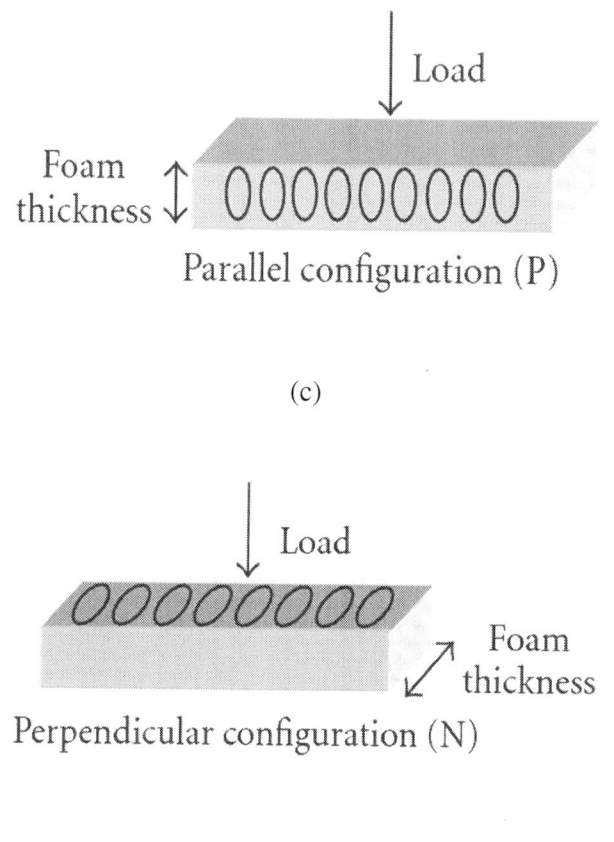

Figure 2: (a) Schematic showing specimen configurations; (b) characteristic cellular structure parameters; (c) parallel (P); (d) perpendicular (N) specimen configurations.

The morphology of the foamed nanocomposites was assessed from high-magnification SEM micrographs and using transmission electron microscopy, TEM (HITACHI H-800). For TEM, sheets with a typical thickness of 60 nm were cut using an ultramicrotome Ultracut E from Reichert-Jung.

Differential scanning calorimetry (DSC) was used to study the thermal characteristics of the matrix. A Perkin Elmer, Pyris 1 model with a glycol-based Perkin Elmer Intracooler IIP calorimeter was employed with samples weighting around 8.0 mg. The following program was used: heating from 30 to 200°C at 10°C/min and holding for 1 min to erase the thermal history, followed by cooling at 10°C/min from 200°C to 30°C and a second heating from 30 to 200°C (10°C/min). The crystallinity percentage (X_c) was determined according to

$$X_c(\%) = \frac{\Delta H_m}{\Delta H_m^0 w_p} \times 100, \tag{1}$$

where W_p is the weight fraction of PP, ΔH_m is the melting enthalpy of the sample and ΔH_m^0 the theoretical, 100% crystalline polypropylene enthalpy (207.1 J/g [27]).

Polypropylene's crystalline characteristics were analyzed by wide angle X-ray scattering (WAXS). A Bruker D8 diffractometer with CuK radiation (λ= 0.154 nm) operating at 45 kV and 40 mA was used. Scans were taken from 1 to 60 with a rotation step of 0.033 and a step time of 0.06 s.

Dynamic mechanical analysis (DMA) was used to measure the dynamic-mechanical properties and study the viscoelastic behaviour of the several foamed nanocomposites. A TA Instruments Q800 Dynamic Mechanical Analyzer was used and calibrated according to the standard procedure. The glass transition temperature (T_g), storage modulus (E'), and loss factor (tanδ) were obtained in a three-point bending configuration using a span length of 50.00 mm. Two configurations, parallel (P) and perpendicular (N), respectively shown in Figures 2(c) and 2(d), were considered. Experiments were performed from −20 to 150°C at 2°C/min and 1 Hz. A static strain of 2% and dynamic of 0.02% with a preload force of 0.01 N and force track of 120% were chosen. Test specimens were prepared in a prismatic shape (see Figures 2(c) and 2(d)) with a nominal length of 55.00 ± 0.10 mm, width of 13.00 ± 0.10 mm, and thickness of 3.00 ± 0.05 mm (solids) and 3.50 ± 0.10 mm (foams). Three experiments were performed for each material. The values reported in the text (T_g, E' and tanδ) are the average of these three experiments, and in all cases the standard deviation was lower than 5%.

The electrical conductivity of the several solid and foamed PP-CNF nanocomposites was measured as a function of frequency between 10^{-2} and 10^6 Hz using a Novocontrol impedance analyzer (HP 4192 A LF). A typical thickness of 130 μm and 1.5 mm was, respectively, used for the solid and foams. All the measurements were made by previously gold-coating the surfaces of the samples, as it has previously been shown to directly affect the measurement of the electrical conductivity [23]. Five experiments were performed for each material, in all cases the standard deviation being lower than 3%.

RESULTS AND DISCUSSION

Foaming Behaviour and Cellular Structure

The material's code, density and respective gas (V_{gas}), polypropylene (V_{pp}) and filler (V_p) volume fractions are presented in Table 1 alongside the most characteristic cellular structure results for all the foams.

Table 1: Cellular characterization results of the unfilled (PP), montmorillonite (PP-MMT), and carbon nanofibre (PP-CNF) polypropylene foams

Code	Foam density (kg·m⁻³)	V_{gas}	V_{PP}	V_p	ϕ_{VD} (µm)	ϕ_{WD} (µm)	AR	Cell density (cells·cm⁻³)
PP	346 ± 9	0.62	0.38	—	194 ± 7	175 ± 1	1.1	4.71 × 10⁴
	311 ± 9	0.66	0.34	—	177 ± 2	196 ± 1	0.9	4.63 × 10⁴
	268 ± 6	0.71	0.29	—	241 ± 9	266 ± 12	0.9	3.52 × 10⁴
	261 ± 13	0.72	0.29	—	243 ± 14	239 ± 2	1.0	3.42 × 10⁴
	236 ± 7	0.75	0.25	—	456 ± 5	479 ± 10	1.0	1.30 × 10⁴
	189 ± 8	0.80	0.20	—	564 ± 6	523 ± 11	1.1	9.13 × 10³

PP-MMT	256 ± 8	0.72	0.28	0.003	146 ± 3	160 ± 7	0.9	9.01 × 10^4
	245 ± 5	0.73	0.27	0.003	169 ± 13	156 ± 12	1.1	8.78 × 10^4
	240 ± 10	0.74	0.26	0.003	190 ± 12	164 ± 15	1.2	6.56 × 10^4
	238 ± 12	0.74	0.25	0.003	296 ± 7	211 ± 4	1.4	3.57 × 10^4
	220 ± 7	0.76	0.24	0.003	218 ± 6	204 ± 8	1.1	4.19 × 10^4
	208 ± 4	0.78	0.22	0.003	289 ± 12	206 ± 6	1.4	3.98 × 10^4
	176 ± 3	0.81	0.19	0.002	410 ± 12	293 ± 12	1.4	1.97 × 10^4
PP-CNF	265 ± 10	0.70	0.29	0.01	569 ± 25	508 ± 19	1.1	8.13 × 10^3
	270 ± 10	0.69	0.29	0.02	462 ± 28	394 ± 22	1.2	1.35 × 10^4
	290 ± 8	0.68	0.29	0.03	239 ± 10	258 ± 12	0.9	3.05 × 10^4

As expected, the average cell size in VD and WD directions increased with foaming (>V_{gas}). Results show that the cell size of the unfilled PP foams increased a lot faster than the PP-MMT ones for V_{gas} between 0.70 and 0.80 (between 250–550 µm for the first and ≈150–300µm for the second, determined as the average of the cell sizes on both direections). The MMT particles reduced the cell sizes (compare Figure 3(a) with Figure 3(c)) and narrowed the cell size distribution (AR closer to 1). The nanoparticles acted as bubble nucleators in the early stages of foaming, locally increasing the melt strength and extensibility of the polymer, thus explaining the increasingly higher cell size differences between both foams.

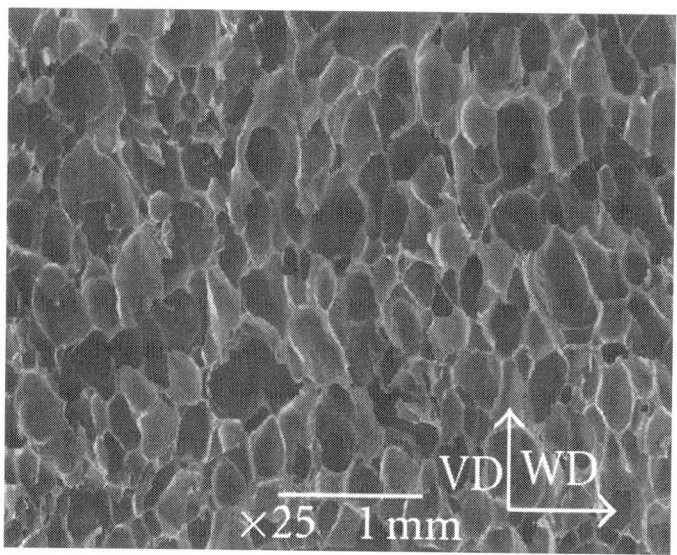

(a)

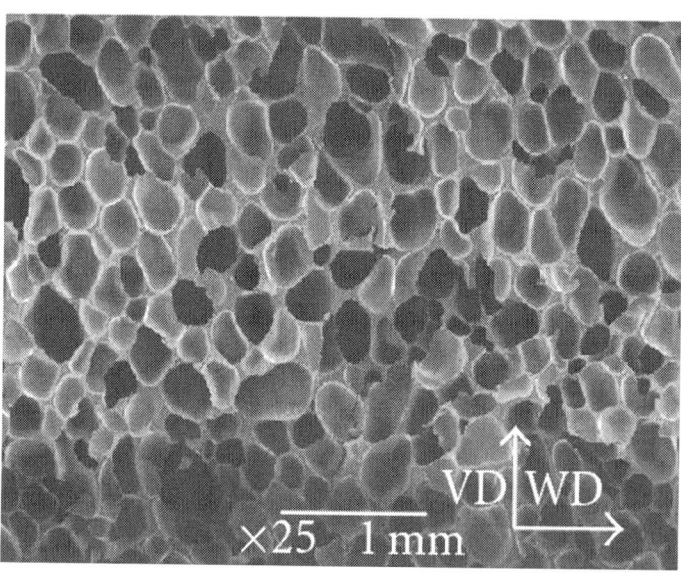

(b)

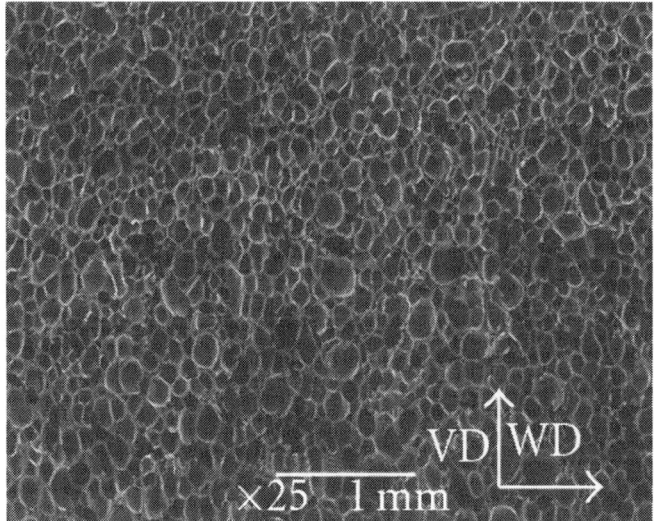

(c)

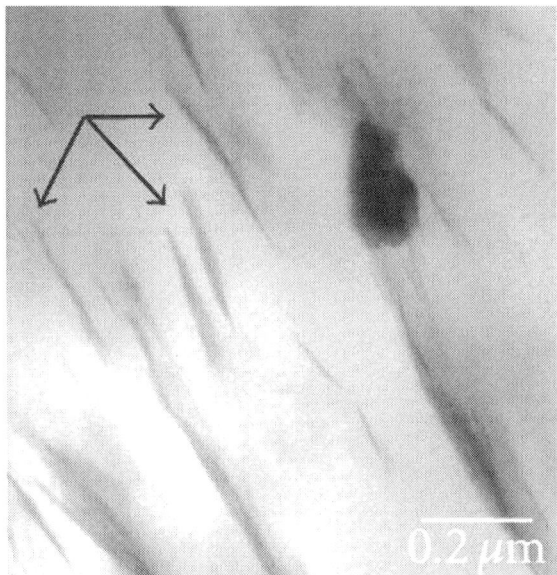

(d)

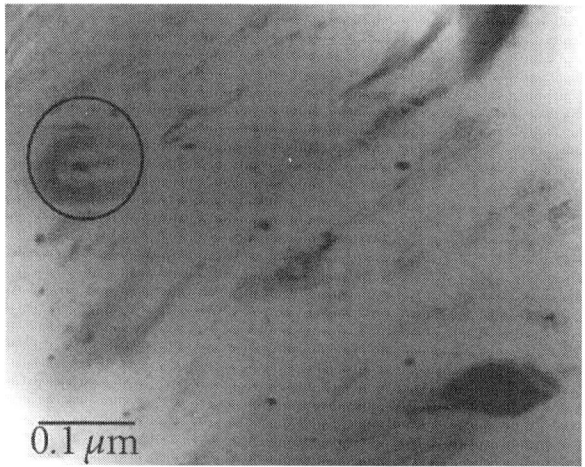

(e)

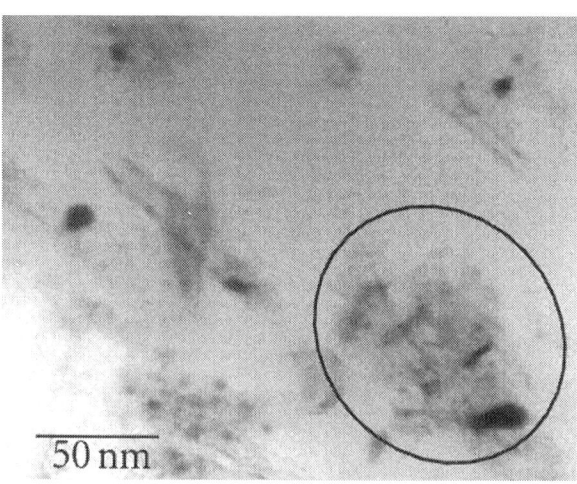

(f)

Figure 3: Typical SEM micrographs (×25) of the (a) unfilled PP, (b) PP-CNF, and (c) PP-MMT nanocomposite foams; (d) typical TEM picture showing partial exfoliation of the MMT platelets; (e) and (f) TEM pictures showing CNF dispersion. VD: Vertical direction of foaming; WD: Width direction.

A good intercalation/exfoliation of the nanoparticles was mainly obtained after foaming the material (see arrows showing partially exfoliated MMT platelets in Figure 3(d) and the WAXS results shown later), thus supporting earlier results stating that foaming could come as a useful tool to exfoliate platelet-like structures such as montmorillonite [6, 16].

In the case of the carbon nanofibre-reinforced foams (PP-CNF), the cell size decreased for similar volume gas fractions with adding increasingly higher amounts of carbon nanofibres (see Table 1). Besides, the carbon nanofibres contributed to the formation of an isotropic cellular structure, that is, foams with aspect ratios close to 1. For similar expansion ratios ($V_{gas} \approx 0.70$), a considerable cell size reduction was observed with increasing the amount of nanofibres from 550 to 250 µm, respectively, for the 10 and 20 wt.% CNF foams. A typical SEM micrograph displaying the cellular structure of the PP-CNF foams is presented in Figure 3(b).

Also presented in Figure 3 are two TEM pictures obtained at different magnifications showing the dispersion of the carbon nanofibres in the PP matrix (Figures 3(e) and 3(f)). Although some aggregates were observed (see black circles), these were a lot scarcer than in similar thermoplastic carbon nanotube-reinforced composites [28], thus supporting the combined efficiency of the melt-mixing and foaming processes.

Effects on the Crystalline Characteristics

The X-ray scattering analysis allowed studying eventual effects induced by the foaming process and incorporation of both types of nanofillers on polypropylene's crystallinity. Typical WAXS spectra of the solid and foamed nanocomposites are shown in Figure 4.

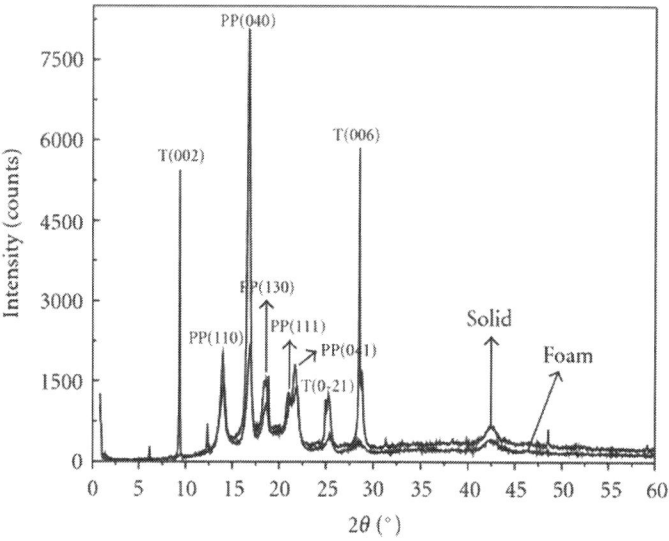

(a)

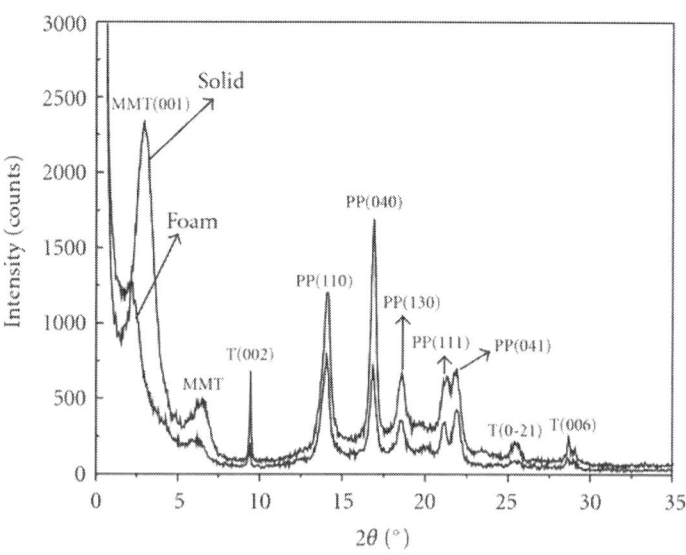

(b)

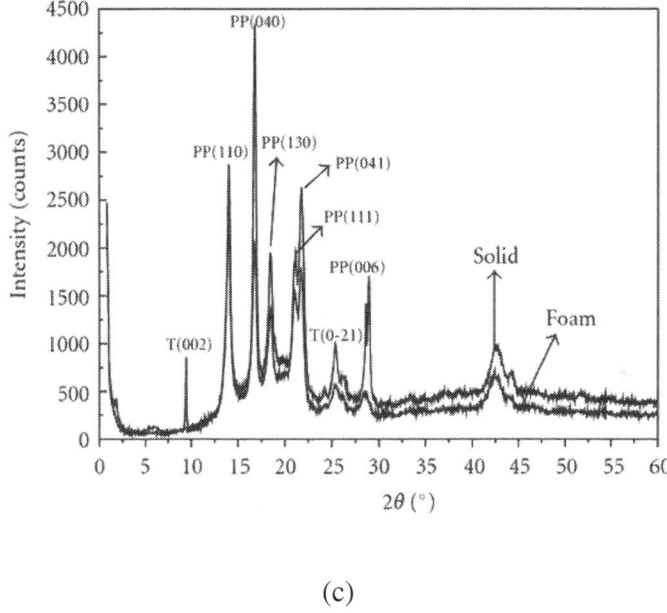

(c)

Figure 4: Typical WAXS spectra of the solid and foamed (a) unfilled PP, (b) PP-MMT and (c) PP-CNF nanocomposites.

A considerable intensity shift from polypropylene's (040) diffraction peak towards the (110) was found with foaming for all the materials. The intensity ratio between peaks, I(040)/I(110), decreased for the unfilled PP from around 4 for the solid to 1.5 for the foam. A remarkable further decrease was found with foaming the nanocomposites: 1.4 to 0.9 and 1.5 to 0.9, respectively, for the MMT and CNF nanocomposite foams. This ratio is directly related to the arrangement of the b lattice parameter of the α-monoclinic polypropylene crystal, a higher value being indicative of a preferential orientation parallel to the sample's surface [29]. Despite the crystal anisotropy induced during the preparation of the solid discs, especially noticeable for the unfilled material (I(040)/I(110) ≈ 4), foaming reduced this preferential crystal orientation. Especially relevant is the fact that foaming totally erased all possible crystal orientation in the case of the nanocomposites (I(040)/I(110) ≈ 1).

Concerning polypropylene's crystallinity, the differences observed between the unfilled materials and the nanocomposites show that there is a less α-crystal perfection in the pure PP foams than that in the solid (lower values of the full width at half maximum, FWHM). Contrarily, the nanoparticles promoted a higher crystalline perfection (lower FWHM values for the foams).

WAXS spectra were also used to ascertain the efficiency of the melt-mixing and foaming processes in guaranteeing an intercalated/exfoliated MMT nanocomposite morphology. Analyzing the (001) MMT peak, a lower diffraction angle was obtained for the foamed nanocomposite compared to its solid counterpart, indicating an increase in the interlayer distance (d_{001}) from 2.99 nm to 4.01 nm. Nevertheless, the foaming process was not enough to promote a total exfoliation of the montmorillonite particles. Despite the decrease in intensity, the (001) peak still appeared for the foamed nanocomposite, as seen in Figure 4(b). In good agreement, TEM analysis showed that the typical morphology of the foamed MMT nanocomposites consisted of mixed dispersed individual montmorillonite and stacks of montmorillonite platelets (see Figure 3(d)).

The PP-MMT nanocomposites showed a higher crystallization temperature and crystallinity than the unfilled polymer measured by DSC, indicating that the well-dispersed nanoparticles acted nucleating crystals. The same type of nucleating effect was observed with incorporating the carbon nanofibres, mainly noticeable for the higher amounts. Higher crystallization temperatures were observed with adding the montmorillonite nanoparticles as well as the carbon nanofibres: from 130.2 and 126.7°C corresponding, respectively, to the beginning and maximum of the crystallization peak of the unfilled PP to 135.3 and 132.9°C for the PP-MMT nanocomposite and 135.6 and 131.8°C for the 20 wt.% CNF material. In the case of the PP-CNF nanocomposites, the crystallization temperature measured at the maximum steadily increased around 2C with adding the nanofibres: from the 126.7°C of PP, to 128.8°C (5 wt.% CNF), 129.6°C (10 wt.% CNF), and 131.8C (20 wt.% CNF). The crystallinity increased from the 46.2% of the material without nanofiller to 49.0% for the PP-MMT nanocomposite and 46.5, 48.9, and 49.6%, respectively, for the 5, 10, and 20 wt% CNF materials. See Figure 5 for comparison between the unfilled and 20 wt% CNF foams.

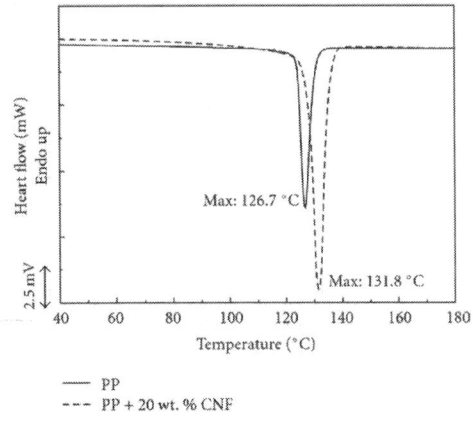

(a)

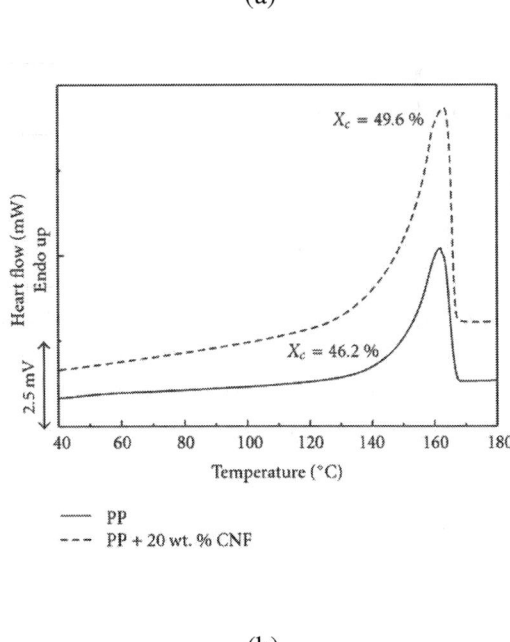

(b)

Figure 5: Comparative differential scanning calorimetry between the unfilled and 20 wt.% CNF PP foams showing (a) the peak's maximum crystallization temperatures and (b) crystallinity determination.

Dynamic-Mechanical Thermal Analysis

Typical DMA curves of the several foamed nanocomposites are presented in the two considered measured directions, that is, parallel (P) and perpendicular (N) to the foam's surface, in Figure 6.

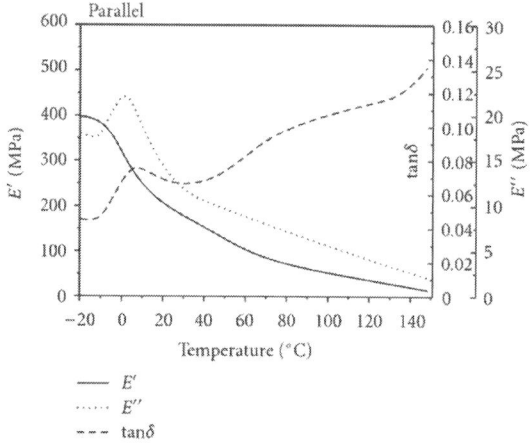

(a)

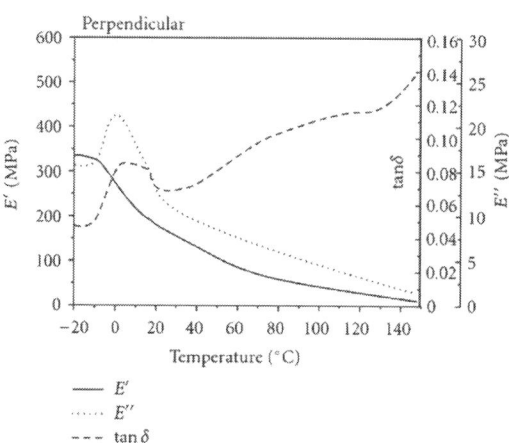

(b)

Foaming Behaviour, Structure, and Properties of Polypropylene... 223

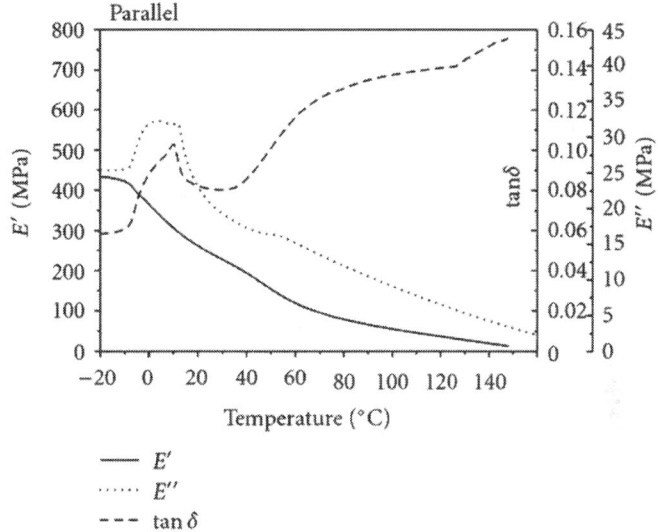

(c)

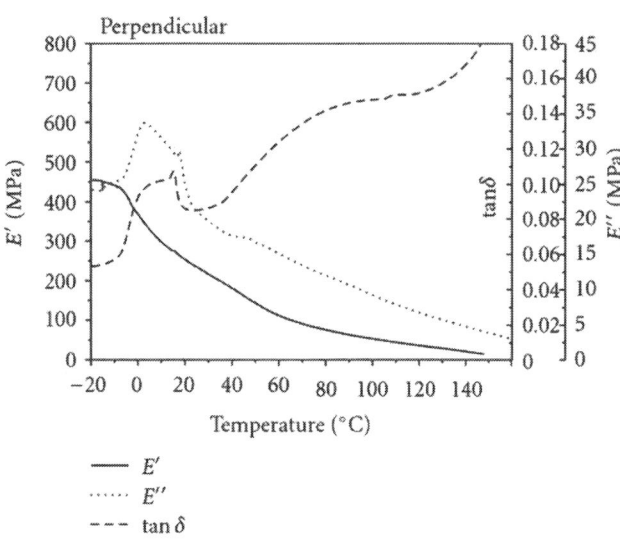

(d)

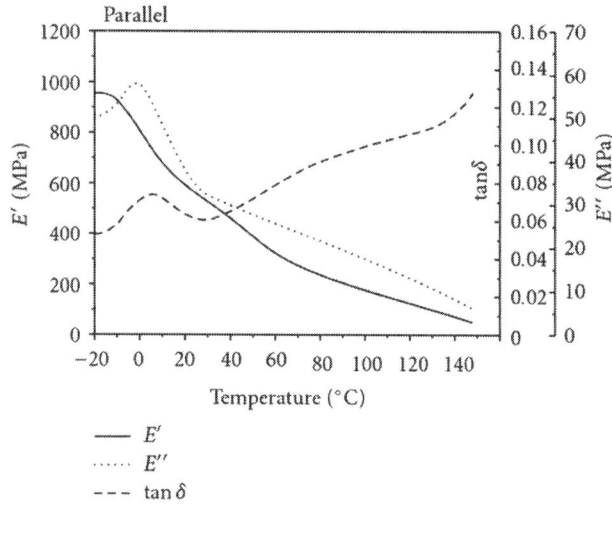

(e)

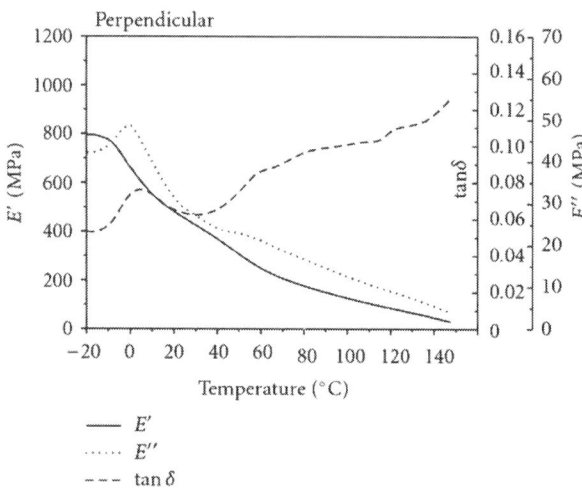

(f)

Figure 6: Typical DMA curves of the unfilled PP foams, (a) and (b); PP-MMT foams, (c) and (d); PP-20% CNF foams, (e) and (f).

At low strain values, the main mechanisms governing the materials response are bending and stretching of the cell walls, thus enabling the study of the viscoelastic relaxations of the polymer matrix. Such is the case of the glass transition temperature (T_g), that accounts for the glass-rubber relaxation of the amorphous portions of the material. Considering that all the foams analyzed here displayed similar expansion ratios, it was possible to analyze the effects of the MMT nanoparticles and carbon nanofibres on the viscoelastic behaviour of the foams.

In the case of the CNF-reinforced nanocomposites, as seen in Table 2, the glass transition temperature (T_g) raised with increasing the amount of nanofibres, from the 5.2°C of the 5 wt.% CNF solid to the 7.3 and 7.5°C, respectively, of the 10 and 20 wt.% CNF solid nanocomposites. This is the direct result of a higher crystallinity (lower amorphous fraction) and less matrix mobility due to the presence of the nanofibres. Nevertheless, its value decreased with foaming for all the materials.

Table 2: DMA results of the solid and foamed unfilled, MMT, and CNF-reinforced PP foams

Material	Direction	T_g*(°C)	E' at 20°C (MPa)	Specific modulus (MPa·cm³·g⁻¹)	S
PP	Solid	5.4	1961.3	2114.4	—
	P	6.6	206.5	676.3	1.1
	N	4.7	179.9	589.2	
PP-MMT	Solid	6.0	2010.0	2152.0	—
	P	5.6	261.5	839.9	1.0
	N	5.9	252.7	823.1	
PP-5% CNF	Solid	5.2	1827.5	2006.0	—
	P	4.5	235.4	801.5	1.1
	N	2.5	209.6	713.6	
PP-10% CNF	Solid	7.3	2300.3	2532.6	—
	P	6.9	231.4	806.6	0.7
	N	5.2	328.6	1100.9	
PP-20% CNF	Solid	7.5	2732.7	2832.1	—
	P	4.5	543.5	1230.0	1.0
	N	4.3	473.6	1185.5	

*T_g—Glass transition temperature measured in tan δ.

As expected, the storage modulus (E') increased with increasing the amount of carbon nanofibres. In the particular case of the foams, the storage modulus only slightly increased with CNF's content for similar relative densities. Nonetheless, the specific storage modulus, that is, the storage modulus relative to the foam's density, increased considerably with adding a higher amount of nanofibres (from the around 800 MPa·cm^3/g of the 5 wt.% CNF foams to the 1200 MPa·cm^3/g of the 20 wt.% CNF ones), indicating the carbon nanofibres efficiency as mechanical reinforcements.

Although displaying almost the same storage modulus than similar relative density unfilled PP foams, comparing the specific value, it is noticeable the reinforcement effect of the montmorillonite particles, with a more than 30% increase.

A parameter S, defined as the specific storage modulus in the parallel direction divided by that in the perpendicular direction was determined as to ascertain the isotropic mechanical properties of the foamed composites. As can be seen by the values presented in the last column of Table 2, the MMT- and CNF-reinforced foams presented a more isotropic mechanical behaviour, especially the PP-MMT nanocomposite foams, displaying an S value of 1, result of their finer isotropic-like cellular structure.

Generally speaking, the MMT nanocomposite foams displayed loss factor values (tan δ) slightly higher than the unfilled ones. Taking into account the high densities of the nanocomposite foams and these less importance of the gas enclosed inside the cells, these slight differences may be attained to the different microstructures of the polymer present in the cell walls, as has been previously shown in the partially exfoliated MMT nanoparticles to induce a higher crystallinity and less crystal anisotropy in the PP.

Electrical Conductivity Measurements

The carbon nanofibres were initially added with the main objective of developing new conductive lightweight materials. With that in mind, measurements of the electrical conductivity of the several solid and foamed PP-CNF nanocomposites were performed over a wide range of frequencies.

Figure 7 presents the broadband electrical conductivity values of the several solid and foamed nanocomposites as a function of frequency.

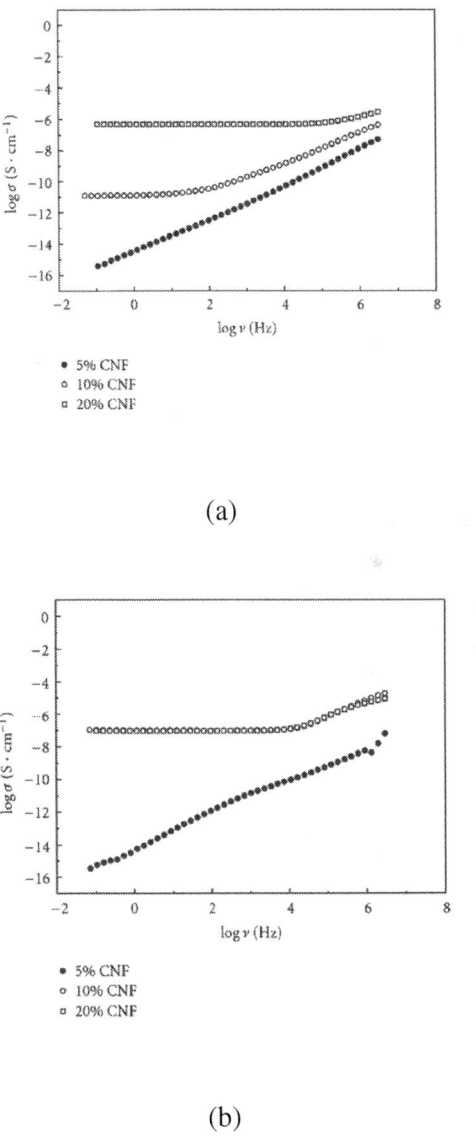

Figure 7: Broad-band electrical conductivity (σ) as a function of frequency (ν) for the (a) solid, and (b) foamed PP-CNF nanocomposites.

In the case of a 5 wt.% CNF content the electrical conductivity of the solids and foams followed a linear behaviour with frequency characteristic of insulating materials such as PP ($\sigma_{PP} \approx 10^{-16}$ S cm^{-1}). This indicates that the electrical properties of the composite are being controlled by the matrix, as the nanofibres are clearly too far apart to allow electrical conduction. Nonetheless, for a 10 wt.% CNF concentration the materials started to show a characteristic electrical conduction behaviour, displaying a critical frequency (v_c) below which conductivity gets frequency independent (known as the direct current conductivity, σ_{dc}). For comparative purposes, the dc conductivity (σ_{dc}) was always taken at the same frequency (10^{-1} Hz).

Comparatively, foams reached a dc conductivity value earlier than the solid materials. For instance, the 10 wt% CNF foam displayed a value of 1.06×10^{-7} S/cm, considerably higher than the 1.22×10^{-11} S/cm of the respective solid. Nevertheless, this conductivity value is still low based on the theoretical value and amount of nanofibres as well as compared to other polymer systems [25]. In these cases, there is a critical concentration of filler, known as the percolation threshold, φ_c, where the formation of a 3D conductive network results in an abrupt increase in the electrical conductivity [30]. It has been shown that the electrical efficiency of these conductive fillers depends on the presence of polymer chains between contacts, and particularly on local crystal formation. Electrical conduction considerably decreases in the crystalline regions compared to the amorphous ones, where ion conductivity is the dominating conduction mechanism [31–33]. These particularities act to the formation of electrical resistances between the nanofibres, limiting an effective electrical conduction by percolation. Under these conditions, a model based on tunnelling conduction fits better to the analyzed system, with the dc conductivity being depicted by $\sigma_{dc} \propto \exp(-Ad)$ [34, 35], where A is a tunnel parameter and d is the so-called tunnel distance. This predominant tunnelling conduction behaviour has been previously shown in other research works to be typical of carbon nanofibre polyolefin nanocomposites, due to fibre breaking during processing the dispersed short carbon nanofibres making it a lot harder to reach a direct contact between the fibres at low contents [36]. Using the tunnelling conduction approach, the theoretical critical concentration of nanofibres for electrical conduction for the nanocomposite foams was found to be 10 wt.% lower than the 12 wt.% determined for the solids, indicating that the foaming process is adding to a higher nanofibre functionality due to improved dispersion (see Figure 3(f)).

CONCLUSIONS

This work presents the preparation and preliminary characterization of montmorillonite and carbon nanofibre-reinforced rigid polypropylene foams, with the objective of developing new multifunctional lightweight materials for structural applications.

Regarding the foaming behaviour and cellular structure of the foams, the incorporation of MMT and CNF resulted in finer isometric-like cellular structures, especially noticeable for the MMT nanocomposite foams. This was attained to a cell nucleation effect, supported by the higher crystallization temperatures and crystallinities. In the case of the PP-CNF foams, increasingly smaller cell sizes were observed for similar expansion ratios with increasing the amount of nanofibres.

The nanofillers, adding to the foaming process, totally erased the preferential crystal orientation of the -PP's crystal observed in the unfilled solid and in a lesser extent in the unfilled PP foams (b lattice parallel to the sample's surface). Although the FWHM values of the unfilled PP increased with foaming, the incorporation of the nanofillers promoted the opposite effect, related to a higher crystalline perfection.

The dynamic-mechanical behaviour of the unfilled and nanocomposite polypropylene foams showed that the nanofillers acted as mechanical reinforcements, increasing the specific storage moduli of the foams. They also promoted a more isotropic-like mechanical behaviour, especially the MMT nanoparticles, related to a finer isotropic cellular structure of the nanocomposite foams.

As a previous step to the study of incorporating conductive nanofillers on the electrical conduction behaviour of PP foams, electrical conductivity measurements were performed on the different solid and foamed PP-CNF nanocomposites. Interesting results were found with foaming the nanocomposites, the foams displaying a higher electrical conductivity than their solid counterparts, indicating that foaming may come as a useful tool in creating an electrically-conductive network.

ACKNOWLEDGMENTS

The financial assistance from the Spanish Ministry of Science and Education for the project MAT2007-62956 is gratefully acknowledged. The authors would like to thank Professor Miguel Mudarra for all the help with the electrical conductivity measurements.

REFERENCES

1. M. A. Rodríguez-Pérez, "Crosslinked polyolefin foams: production, structure, properties, and applications," Advances in Polymer Science, vol. 184, pp. 97–126, 2005.
2. C. B. Park and L. K. Cheung, "A study of cell nucleation in the extrusion of polypropylene foams,"Polymer Engineering and Science, vol. 37, no. 1, pp. 1–10, 1997.
3. Sekisui Alveo, Product Guide, Sekisui Alveo, Roermond, The Netherlands, 2003.
4. S. Tokuda and T. Kemmotsu, "Electron beam irradiation conditions and foam seat properties in polypropylene-polyethylene blends," Radiation Physics and Chemistry, vol. 46, no. 4–6, pp. 905–908, 1995.
5. P. Spitael, C. W. Macosko, and A. Sahnoune, "Extensional rheology of polypropylene and its effect on foaming of thermoplastic elastomers," in Proceedings of the Annual Technical Conference (ANTEC '02), pp. 1792–1796, Society of Plastics Engineers, San Francisco, Calif, USA, 2002.
6. M. Antunes, J. I. Velasco, V. Realinho, and E. Solórzano, "Study of the cellular structure heterogeneity and anisotropy of polypropylene and polypropylene nanocomposite foams," Polymer Engineering and Science, vol. 49, no. 12, pp. 2400–2413, 2009.
7. N. Mills, Polymer Foams Handbook. Engineering and Biomechanics Applications and Design Guide, Elsevier, Oxford, UK, 1st edition, 2007.
8. D. Klempner and V. Sendijarevic, Polymeric Foams and Foam Technology, Hanser, Munich, Germany, 2nd edition, 2004.

9. R. R. Puri and K. T. Collington, "The production of cellular crosslinked polyolefins. 2. The injection-molding and press molding techniques," Cellular Polymer, vol. 7, pp. 219–231, 1988.
10. UK Zotefoams. High Performance Polymers, October 1999.
11. D. Eaves, Handbook of Polymer Foams, Rapra Technology, Shawbury, UK, 2004.
12. M. Antunes, V. Realinho, and J. I. Velasco, "Study of the influence of the pressure drop rate on the foaming behaviour and dynamic-mechanical properties of CO_2 dissolution microcellular polypropylene foams," Journal of Cellular Plastics. In press.
13. J. I. Velasco, M. Antunes, O. Ayyad, et al., "Foaming behaviour and cellular structure of LDPE/hectorite nanocomposites," Polymer, vol. 48, no. 7, pp. 2098–2108, 2007.
14. J. E. Martini, N. P. Suh, and F. A. Waldman, US patent 4,473,665, 1984.
15. L. J. Lee, C. Zeng, X. Cao, X. Han, J. Shen, and G. Xu, "Polymer nanocomposite foams," Composites Science and Technology, vol. 65, pp. 2344–2363, 2005.
16. C. Zeng, X. Han, L. J. Lee, K. W. Koelling, and D. L. Tomasko, "Polymer-clay nanocomposite foams prepared using carbon dioxide," Advanced Materials, vol. 15, no. 20, pp. 1743–1747, 2003.
17. Y. Di, S. Iannace, E. Di Maio, and L. Nicolais, "Poly(lactic acid)/organoclay nanocomposites: thermal, rheological properties and foam processing," Journal of Polymer Science—Part B, vol. 43, no. 6, pp. 689–698, 2005.
18. X. Han, C. Zeng, L. J. Lee, K. W. Koelling, and D. L. Tomasko, "Extrusion of polystyrene nanocomposite foams with supercritical CO_2," Polymer Engineering and Science, vol. 43, no. 6, pp. 1261–1275, 2003.
19. M. Okamoto, P. H. Nam, P. Maiti, et al., "Biaxial flow-induced alignment of silicate layers in polypropylene/clay nanocomposite foam," Nano Letters, vol. 1, no. 9, pp. 503–505, 2001.
20. M. Lee, B.-K. Lee, and K.-D. Choi, "Foam compositions of polyvinyl chloride nanocomposites," WO/2004/074357, International Application No.: PCT/KR2004/000328. 2004.

21. J. Shen, X. Han, and L. J. Lee, "Nucleation and reinforcement of carbon nanofibers on polystyrene nancomposite foam," in Proceedings of the 63rd Annual Technical Conference (ANTEC '05), vol. 7, pp. 117–121, November 2005.
22. M. Antunes, V. Realinho, A. B. Martínez, E. Solórzano, M. A. Rodríguez-Pérez, and J. I. Velasco, "Heat transfer of mineral-filled polypropylene foams," Defect and Diffusion Forum, vol. 297–301, pp. 990–995, 2010.
23. M. Antunes, J. I. Velasco, V. Realinho, and D. Arencón, "Characterization of carbon nanofibre-reinforced polypropylene foams," Journal of Nanoscience and Nanotechnology, vol. 10, pp. 1241–1250, 2010.
24. M. Antunes, V. Realinho, E. Solórzano, M. A. Rodríguez-Pérez, J. A. de Saja, and J. I. Velasco, "Thermal conductivity of carbon nanofibre-polypropylene composite foams," Defect and Diffusion Forum, vol. 297–301, pp. 996–1001, 2010.
25. M. Shaffer and J. Sandler, "Carbon nanotube/nanofibre polymer composites," in Processing and Properties of Nanocomposites, S. Advani, Ed., pp. 1–59, World Scientific, River Edge, NJ, USA, 2006.
26. G. L. A. Sims and C. Khunniteekool, "Cell size measurement of polymeric foams," Cellular Polymers, vol. 13, no. 2, pp. 137–146, 1994.
27. B. Wunderlich, Thermal Analysis, Academic Press, New York, NY, USA, 1990.
28. J. Sandler, G. Broza, M. Nolte, K. Schulte, Y.-M. Lam, and M. S. P. Shaffer, "Crystallization of carbon nanotube and nanofiber polypropylene composites," Journal of Macromolecular Science, vol. 42, no. 3-4, pp. 479–488, 2003.
29. J. I. Velasco, C. Morhain, A. B. Martínez, M. A. Rodríguez-Pérez, and J. A. de Saja, "The effect of filler type, morphology and coating on the anisotropy and microstructure heterogeneity of injection-moulded discs of polypropylene filled with aluminium and magnesium hydroxides. Part 1. A wide-angle X-ray diffraction study," Polymer, vol. 43, no. 25, pp. 6805–6811, 2002.
30. D. Stauffer and A. Aharony, Introduction to Percolation Theory, Taylor & Francis, London, UK, 2nd edition, 2003.

31. P. Pötschke, M. Abdel-Goad, I. Alig, S. Dudkin, and D. Lellinger, "Rheological and dielectrical characterization of melt mixed polycarbonate-multiwalled carbon nanotube composites," Polymer, vol. 45, no. 26, pp. 8863–8870, 2004.
32. I. Alig, S. M. Dudkin, W. Jenninger, and M. Marzantowicz, "Ac conductivity and dielectric permittivity of poly(ethylene glycol) during crystallization: percolation picture," Polymer, vol. 47, p. 1722, 2006.
33. I. Alig, D. Lellinger, S. M. Dudkin, and P. Pötschke, "Conductivity spectroscopy on melt processed polypropylene-multiwalled carbon nanotube composites: recovery after shear and crystallization," Polymer, vol. 48, no. 4, pp. 1020–1029, 2007.
34. E. K. Sichel, J. I. Gittleman, and P. Sheng, "Tunneling conduction in carbon-polymer composites," inCarbon Black Polymer Composites: The Physics of Electrically Conducting Composites, E. K. Sichel, Ed., Marcel Dekker, New York, NY, USA, 1982.
35. N. Ryvkina, I. Tchmutin, J. Vilčáková, M. Pelíšková, and P. Sáha, "The deformation behavior of conductivity in composites where charge carrier transport is by tunneling: theoretical modeling and experimental results," Synthetic Metals, vol. 148, no. 2, pp. 141–146, 2005.
36. A. Linares, J. C. Canalda, M. E. Cagiao, et al., "Broad-band electrical conductivity of high density polyethylene nanocomposites with carbon nanoadditives: multiwall carbon nanotubes and carbon nanofibers," Macromolecules, vol. 41, no. 19, pp. 7090–7097, 2008.

Chapter 8

Natural Products: A Minefield of Biomaterials

Oladeji O. Ige[1] Lasisi E. Umoru[1], and Sunday Aribo[2]

[1]Department of Materials Science and Engineering, Obafemi Awolowo University, Ile-Ife 220282, Nigeria

[2]Department of Metallurgical and Materials Engineering, Federal University of Technology, Akure 340252, Nigeria

ABSTRACT

The development of natural biomaterials is not regarded as a new area of science, but has existed for centuries. The use of natural products as a biomaterial is currently undergoing a renaissance in the biomedical field. The major limitations of natural biomaterials are due to the immunogenic response that can occur following implantation and the lot-to-lot variability in molecular structure associated with animal sourcing. The chemical stability and biocompatibility of natural products in the body greatly accounts for their utilization in recent times. The paper succinctly defines biomaterials in terms of natural products and also that natural products as materials in biomedical fields are considerably versatile and promising. The various types of natural products and forms of biomaterials are highlighted. Three main areas of applications of natural products as materials in medicine are described, namely, wound management products, drug delivery systems, and tissue engineering. This paper presents a brief history of natural products as biomaterials, various types of natural biomaterials, properties, demand and economic importance, and the area of application of natural biomaterials in recent times.

INTRODUCTION

A biomaterial is regarded as any nondrug material that can be used to treat, enhance or replace any tissue, organ, or function in an organism [1]. While the definition of biomaterial was reframed as a nondrug substance suitable for inclusion in systems which augment or replace the function of bodily tissues or organs [2] This definition explicitly described biomaterial in relation to drugs and as such, there is a need to clarify the impression that natural products are synonymous with drugs. The definition implies that natural products can be applied as biomaterials by eliminating the ambiguity always associated with natural products as drugs. It must be emphasized that this definition is not regarded as one of the most popular and is not often cited as this one which defines biomaterial as a nonviable material that intends to interact with physiological environment [3]. However, in this study, the following definition will be adopted: biomaterial can be defined as any substance (other than a drug) or combination of substances synthetic or natural in origin, which can be used any time, as a whole or as a part of a system which treats, augments, or replaces any tissue, organ or function of the body [4]. It must be noted that in this study the substances are natural in origin.

Economy

The field of biomaterials working under biological constraints is rapidly expanding. It is considered that this domain represents 2-3% of the overall health expenses in developed countries [5]. There are over 13 million medical device implants annually in USA. However, all these devices are prone to incomplete or nonspecific cellular healing which may lead to the ultimate failure of the device [6].

History

The new interest in natural biomaterials could really be considered as a renaissance. Historians have traced the use of sutures made from animal sinew to ancient Egypt, while some say they were used even earlier [7]. The ancient Egyptians were regarded as the first to use biomedical materials, employing coconut shells to repair injured skulls; wood and ivory as false teeth and these dated as far back as 3000 BC [8]. Some of the earliest biomaterial applications were as far back as ancient Phoenicia where loose teeth were bound together with gold wires, tying artificial teeth to neighbouring teeth [2].

As early as the first century AD in both Greece and India, physicians were using natural biomaterials while performing plastic surgery to repair mutilations from battle and punishment. There are even accounts of some of these physicians treating disemboweled soldiers to good effect [7]. The earliest recorded hip replacement surgery took place in Germany in 1891 AD, and in this instance ivory was used. This material is favoured for its relatively inexpensive cost although it also turned out to have useful biomechanical bonding qualities, which makes it well suited to work with human body tissue [8]. From as early as a century ago artificial, materials and devices have been developed to a point where they can replace various components of the human body. These materials are capable of being in contact with body fluids and tissues for prolonged periods of time, whilst eliciting little if any adverse reactions occur [9].

In the early 1900's, bone plates were successfully implemented to stabilize bone fractures and to accelerate their healing. While by the time 1950's and 60's, blood vessel replacement was in clinical trials artificial heart valves and hip joint were in development [2]. It is believed that by knowing the history of the development of biomaterials we can have a better perspective of how the state of the art is progressing. The big question with biomaterials is how will this product perform better than those available today? It is all too easy when performing research in today's compartmentalized and specialised society to forge the bigger picture. In the past, the research approach to biomaterials was spurred on by innovative and creative doctors who used whatever they thought might work to fix a problem sometimes with wonderful results but with extremely high risk. Today's researchers are much more focused into improving medicine methodically and with as little risk to patients as possible. The road has been long and treacherous; nature has given us many amazing remedies put together by millions of years of evolution [10].

Theoretically, any material natural or manmade can be a biomaterial as long as it serves the stated medical and surgical purposes. The development of biomaterials is not a new area. It encompasses elements of medicine, biology, chemistry, tissue engineering, and materials science. Nevertheless, the demand for biocompatible, biodegradable, and bioresorbable materials has increased dramatically during the last decade. An ideal biomaterial is one that is nonimmunogenic, biocompatible, and biodegradable, which can be functionalized with bioactive proteins and chemicals. In particular, biodegradability is one of the essential properties of the biomaterials [11]. It must be emphasized that the key factors in biomaterial usage are its biocompatibility, biofunctionality, and availability to a lesser extent.

Biomaterials are ultimately intended for implantation in or on the human body. They can be designed with a range of properties that are capable of either promoting or inhibiting specific host cell and tissue responses [12]. Biomaterials also refer to biologically-derived materials used for their structural rather than biological properties, for example, collagen (a protein found in the skin, connective tissues, and bone) as a cosmetics ingredient. Also carbohydrates (biotechnologically modified) are being used as lubricants for biomedical applications and as bulking agents in the food industry [1]. Not until recent times have naturally derived biomaterials been explored as facilitators and promoters of healing and regeneration. Today, biomaterials of all types are being used for everything from wound dressing to tendon and ligament repair. Extensive experimentation has been undertaken to identify the composition, mechanical properties, and in vivo response of naturally derived biomaterials. The choice of biomaterials depends on the type of procedure being performed, the severity of the patient's condition, and the surgeon's preference. To be successful, the implant should effectively repair the defect it covers without eliciting an adverse tissue reaction while maintaining mechanical and biological integrity for a desired amount of time from a few weeks to several years. The prime reason biomaterials have come about is to provide a remedy for surgical problems [10].

Biomaterials research is one of the most important fields of modern medicine. Biomaterials are used in organ implants, wound healings, drug delivery, and so forth. It was suggested that for many reasons natural biomaterials are the most preferred ones; they are biodegradable, biocompatible, and nontoxic [13]. Biomaterials can be divided into four major classes of materials, namely; polymers, metals, ceramics (including carbons, glass-ceramics, and glasses), and natural materials (including those from both plants and animals). Composite materials are the ones which are comprised of two or more different classes of materials and they are regarded as the fifth class of biomaterials [14]. It is observed that metals rarely occur in nature as a single entity; hence, they are synthesized from their compound ores such as oxides, sulphides, and carbonates with the exception of rare metals like gold, and platinum. Natural metals have found application as dental materials in ancient times. Recently, the applications of metals as natural biomaterials are not documented. There is an international resurgence of interest in natural products as a source of novel bioactive substances for the development of novel drugs and therapies, the same trend is observed in the application of natural products as materials in medicine. By 2020, Ireland will have a leading capacity in the utilization of natural biomaterials and nutraceuticals [15].

There are several materials derived from the animal or plant world being considered for use as biomaterials and they are called natural biomaterials. One of the advantages of using natural materials for implants is that they are similar to materials, and are familiar to the body systems. In this regard, the field of biomimetics (or mimicking nature) is growing. However, natural materials can be subjected to problems of immunogenicity. Another problem faced by these materials, essentially natural polymers, is their tendency to denature or decompose at temperatures below their melting point. This severely limits their fabrication into implants of different sizes and shapes.

Natural materials do not usually offer the problems of toxicity often faced by synthetic materials. Also, they may carry specific protein binding sites and other biochemical signals that may assist in tissue healing or integration [16]. This paper intends to highlight the different types of natural products that are currently being employed as biomaterials, state their advantages and disadvantages, and also to summarize the areas of application in the medical industry.

NATURAL PRODUCTS

Biological structures have always been a source of inspiration for solving technical challenges in architecture, mechanical engineering, or materials science. Nature has developed—with comparatively few base substances, mainly polymers and minerals—a range of materials with remarkable functional properties [17]. Natural was defined as something that is present in or produced by nature and not artificial or man-made and most often the definition is assumed to mean something good or pure [18].

The term natural products is quite commonly understood to refer to herbs, herbal concoctions, dietary supplements, traditional medicine, or alternative medicine [19]. It must be stated that while the stories of herbs and drugs are very much intertwined, it needs to be fully appreciated that the use of herbs as natural product therapy is different from their use on the platform of drug discovery and further development [20]. Natural products are regarded as chemicals, but they are not just accidents or products of convenience of nature. More than likely, they are a natural expression of the increase in complexity of organisms [21]. There are several definitions of natural products, and the common trend is that a natural product is a chemical compound or substance produced by a living organism. The living organism is regarded to be found in nature, and that they have pharmacological or biological activity for use in pharmaceutical drug discovery and drug design. A natural product can be considered as such, even if it can be prepared by total synthesis. It must be recognized that not all natural products can be fully synthesized, and many natural products have very complex structures that are too difficult and expensive to synthesize on an industrial scale. Such compounds can only be harvested from natural sources—a process which can be tedious, time-consuming, and expensive as well as being wasteful of the natural resource [22, 23]. It will be realised that natural products are regarded as only beneficial to the pharmaceutical industry at the expense of others like agriculture, food and even the chemical industry. Hence, there is a need to redefine natural products, and according to the biology dictionary definition, natural products depend on the industry; medicine and pharmaceuticals; healthcare and nutritional supplement; agrochemical; food and flavouring. In this study, natural products are regarded as chemical substances that are produced by living organisms with the intent of their application in/as biomedical materials.

It must be noted that interest in natural sources to provide treatments for pain, palliatives, or curatives for a variety of maladies or recreational use dates back to the earliest parts of history. Also, several sources of information on natural products such as pharmaceutical products are well recorded and documented [20]. There are numerous studies on the applications of natural products as biomaterials, but a comprehensive summary has not been reported. Hence this paper intends to give a comprehensive review of natural product usage as biomedical materials.

Natural products can come from anywhere; generally, they are either of prebiotic origin or originate from microbes, plants/animal sources [24]. Suffice it to say that natural products can come from any point or level on the phylogenetic tree. When searching for natural products, one should never feel that a form of life is too low, simple or grotesque to provide a compound of interest. However, before one goes marching out into the woods, sailing out into the sea, climbing the highest mountains, or descending into the deepest caves, it is appropriate to perform a little bit of research, and hence a visit to the library becomes the first step in any search for a natural product. Ideally, it is important to know the history, folklore, origin of use, source, chemical structure, availability, and method of preparation, pharmacology, toxicology, and therapeutics of any natural products [20]. It is observed that it can be challenging to obtain information from practitioners of traditional medicine—which is always well versed in natural products—unless a genuine long-term relationship is made. Natural products may be extracted from tissues of terrestrial plants, marine organisms, or microorganism fermentation broths. A crude extract from any one of these sources typically contains novel, structurally diverse chemical compounds. The sources of natural products are as outlined after which a brief introduction of biomaterials will come up in the next section.

It is noted that several opportunities abound in nature that can provide natural products, and these opportunities can present themselves from almost any niche of nature and most likely some that have not even yet been discovered.

Microorganisms have proven to be an excellent source of natural products including polyketide and peptide antibiotics as well as classes of other biological active compounds [25]. It is noteworthy that some of these compounds when originally discovered failed in their development and their original uses as either antibiotics or as agricultural fungicides [20]. Microbes can be of any sources be it aerial, terrestrial, or even marine. It has been well reported in a wide range of applications in drug synthesis and design. Also, it is used in the biomedical industry as materials for wound management and drug delivery system among others. The diversity of microorganisms is of a staggering quantity, and only an extremely small proportion of bacteria and fungi have been examined for the production of potentially useful secondary metabolites. Bacteria, smuts, nests, yeasts, moulds, fungi, and many other forms of what we consider to be primitive life can be very useful [20].

People most commonly think of plants first when talking about natural products, but trees and shrubs can also provide excellent sources of biomaterials. Plants produce a variety of different types of compounds including biologically active proteins. Some of these types of compounds are even shared with other organisms, and they include such chemical families as lectins, defensins, cyclotides, and ribosome-inactivating proteins [25]. Ribosome-inactivating proteins are a group of proteins exhibiting a wide spectrum of biological activities, including a ribonucleolytic activity for which the group is named. Plant antimicrobial peptides comprise another large group of biologically active compounds. This group of compounds can be further subdivided into thionins, defensins, cyclotides, and lectins. Thionins are small proteins that selectively form disulfides bridges with other proteins or form ion channels in membranes. This ability to make membranes more permeable suggests the potential for antimicrobial activity. Defensins are cysteine-rich peptides that also permeabilize membranes but appear to be very specific in their activity. Finally, lectins are proteins that have a noncatalytic domain that binds reversibly to specific carbohydrates; this activity encompasses potentially a wide spectrum of biological activities. Plants have been, over time, an extremely popular source of natural products [26–29]. Compounds isolated and identified from this source will undoubtedly continue to make strong contributions. Plants as biomedical materials have found applications in wound management, drug delivery systems, and medical fibres and textiles.

Various polypeptides of interest have been isolated from the venom of arachnids and anthropods that prey on insects [25]. Insect peptides have been the subject of research into the immune defense of insects but have not yet been investigated for effects and potential benefit in humans. Compounds from this peptide group include such sources as the termite (Pseudacanthotermes Spiniger), the mosquito (Anopheles gambiae), the moth (Heliothis virescens) and the beetle (Oryctes rhinoceros) [20]. Insect derived natural products offer another strong potential avenue for the development of future biomaterials. However, silk produced by silkworm, Bombyx mori, has excellent properties such as biocompatibility, biodegradation, nontoxicity, and adsorption properties. It has been commercially used as a biomaterial suture for decade and its application also includes wound management, enzyme immobilization matrices, vascular prostheses, structural implants among others [11, 30–35].

Animals, whether highly developed or poorly developed, whether they live on land, in the sea, or in the air can be excellent sources of natural products [20]. Research into a variety of antimicrobial peptides, such as megamins, defensins, cathelicidins, and protegrins generated by vertebrates has, over recent time, become popular. Cathelicidins-type peptides are a broad range of antimicrobial proteins that have been isolated from rabbits, mice, sheep, and humans. Some compounds have been the subject of substantial research, and they offer potential opportunities in the areas of cardiovascular function, immune and central nervous system functions. A typical application is in the area of tissue engineering and these include the use of tissue replacement from animals, dead corpses, and the man himself. These methods are known as xenograft, autograft, and allograft respectively.

The first discovery of a marine-based biologically active compound of interest was really quite by accident approximately 10 years after the end of World War II [36]. The marine environment, arguably the original source of all life, is a rich source of bioactive compounds [37–43]. More than 70% of our planet's surface is covered by oceans, and some experts feel that the potentially available biodiversity on the deep sea floor or coral reefs are greater than those existing in the rainforests [44]. The search for new biomedicals from marine organisms resulted in the isolation of more or less 10,000 metabolites with a broad spectrum of biological activities [45, 46]. There are wide applications of natural products obtained from the marine world such as biomedical materials. A typical example is the successful use of natural coral as a bone graft substitute in tissue engineering.

BIOCERAMICS

It has been accepted that no foreign material placed within a living body is completely compatible. The only substances that conform are those manufactured by the body itself (autogeneous) and any other substances are recognized as foreign, they initiate some type of reaction (host tissue response). Bioceramics are classified according to their bioactivity, and they are, namely, bioinert (alumina dental implant), bioactive [hydroxyapatite, $Ca(PO_4)_2$], surface active (bioglass), and bioresorbable [tricalcium phosphate implant, $Ca_3(PO_4)_2$] [2].

In the early 70's, bioceramics were employed to perform singular biologically inert roles, such as to provide parts for bone replacement. The demands of bioceramics have changed from maintaining an essentially physical function without eliciting a host response to providing a more integrated interaction with the host. Bioceramics potentially can be used as body interactive materials, helping the body to heal, or promoting regeneration of tissues, thus restoring physiological functions. Ultimately, the field of bioceramics is fundamental to advances in the performance and function of medical devices and is a critical part of medicine and surgery. The correlations between material properties and biological performance will be useful in the design of improved bioceramics, particularly to overcome the problems of implant rejection and related infection [2]. Bioceramics have been proposed for biomedical applications such as dental restorations, middle ear reconstruction, rebuilding of facial and cranial bones, and filling of bony defects, to name a few [2, 47]. It was reported that commercially available porous bioceramics originate from two sources, namely; hydroxyapatite (e.g., Pro osteon) or bone (e.g., Endobon) [48].

In this paper, bioceramics that are sourced from natural products are to be considered and there are potentials in shells, coral, bone and soil/mineral.

Bone

Bone consists of a compact tissue type (cortical bone) and a spongy porous material (trabecullar bone). In both tissue types, the basic building block is a bone lamella, typically about 5 μm thick. In cortical bone, lamellae form laminated cylindrical composite structures built around blood vessels, which are denoted as secondary osteons [49]. The mechanical performance of bone, often coined bone quality, does not only depend on the shape and the amount of the bone (as estimated by the bone mineral density, BMD), but also on its architecture and on the quality of the bone material [50, 51].

Bone can be utilized as a biomedical material when it is used to substitute a damaged part. The grafting involving the use of the patient's own bone in replacing the fractured part is known as autografting, whileallografting is the use of another human being's bone and often it involved the utilization of cadaver. Lastly, when animal bones such as rabbit, pig, and dog among others are used as replacement this is known asxenografting [14].

Bone typically consists by weight of 25% water, 15% organic materials and 60% mineral phases. The mineral phase consists primarily of calcium and phosphates ions, with traces of magnesium, carbonate, hydroxyl, chloride, fluoride, and citrate ions. Hence, calcium phosphates occur naturally in the body, but they occur also within nature as mineral rocks, and certain compounds can be synthesized in the laboratory. Also, certain compounds are useful for implantation in the body since both their solubility and speed of hydrolysis increase with decreasing calcium-to-phosphorus ratio [48]. Table 1 summarizes the mineral name, chemical name, and compositions of various phases of calcium phosphates. Driessens stated that those compounds with a Ca/P ratio of less than 1:1 are not suitable for biological application [52].

Table 1: Chemical composition of some calcium phosphates minerals

Ca:P	Mineral name	Formula	Chemical use
1.0	Monetite	$CaHPO_4$	Dicalcium phosphate (DCP)
1.0	Brushite	$CaHPO_4 \cdot 2H_2O$	Dicalcium phosphate dihydrate (DCPD)
1.33	—	$Ca_8(HPO_4)_2(PO_4)_4 \cdot 5H_2O$	Octacalcium phosphate (OCP)
1.43	Whitlockisite	$Ca_{10}(HPO_4)(PO_4)_6$	—
1.5	—	$Ca_3(PO_4)_2$	Tricalcium phosphate (TCP)
1.67	Hydroxyapatite	$Ca_{10}(PO_4)_6(OH)_2$	—
2.0	—	$Ca_4P_2O_9$	Tetracalcium phosphate

The main crystalline component of the mineral phase of bone is a calcium-deficient carbonate hydroxyapatite.

Hydroxyapatite

This has been a well-known biomaterial since the 1960's, and it is universally accepted for the excellent biocompatibility, nondegradability, good haemocompatibility, noncarcinogenic, and nonimmunogenic reaction. It also releases calcium and phosphate ions, and it leads to osteoinduction. Hydroxyapatites are inorganic materials and based on their properties that are widely used in tissue engineering and bone replacement, it is also used as a coating for fixing prostheses [53, 54]. Hydroxyapatite has been used clinically in a wide range of forms and applications. However, one of the major contributions of clinical application of hydroxyapatite is in the form of filler in a polymer matrix [48]. The natural sources of hydroxyapatite are regarded to be safer due to their cross-reaction and other immunological reaction as compared to synthetic hydroxyapatite [54].

Coral

Natural coral (Porites) consists of a mineral phase principally calcium carbonate in the structural form of aragonite with impurities such as strontium, magnesium and fluoride ions, and an organic matrix.

Commercially available coral (Biocoral) is used as bone graft material and has been reported to be biocompatible and resorbable [55, 56]. Natural coral graft substitutes are derived from the exoskeleton of marine madreporic corals. Researchers first started evaluating coral as a potential bone graft substitute in the early 1970s in animals and in 1979 in humans. The structure of the commonly used coral Porite, is similar to that of cancellous bone, and its initial mechanical properties resemble those of bone. The exoskeleton of these high content calcium carbonate scaffolds has since been shown to be biocompatible, osteoconductive, and biodegradable at variable rates depending on the exoskeleton porosity, the implantation site and the species. Although not osteoinductive or osteogenic, coral grafts act as an adequate carrier for growth factors and allow cell attachment, growth spreading, and differentiation. When applied appropriately and when selected to match the resorption rate with the bone formation rate of the implantation site, natural coral exoskeletons have been found to be impressive bone graft substitutes [57]. There are several natural sources of coral and coral-derived materials, and they are resorbed slowly and substituted by host bone. Only two of the naturally occurring coral-based materials will be discussed.

Coralline apatites can be derived from sea coral and they are a naturally occurring structure with optimal strength and structural characteristics. It is known that the pores structure of coralline calcium phosphate produced by certain species is similar to human cancellous bone, making it suitable material for bone graft applications. Coral and converted coralline hydroxyapatite have been used as bone grafts and orbital implants since the 1980s, as the porous nature of the structure allows ingrowth of blood vessels to supply blood to the bone, which eventually infiltrates the implant. Several methods have been developed in producing hydroxyapatite directly from corals but two of such processes are widely utilized and reported. These techniques are hydrothermal and microwave processes, although others such as sol-gel coating are also in use among others [53].

Coral sands have the potential for a range of biomedical applications where calcium phosphates are used. There is difficulty associated with manufacturing of spherical porous powders which has increased the interest in discovering possible alternatives in nature. It was reported that under controlled hydrothermal exchange, a mix product of calcite and β-tricalcium phosphate can be derived from coral sand grains, and they were able to retain their porous structures. This makes them suitable for potential biomedical application as biodegradable material [58].

Corals, including those of Indian and Australian origin, have been converted to coralline hydroxyapatite successfully in the past [59, 60]. Considerable efforts are needed in order to consider the possibilities of converting African corals to coralline hydroxyapatite especially in the west coast of the continent where these substances are present in abundance. Coral has architectural properties such as interconnected macroporosity, which can be adapted to the biological requirements of the receiver bone. Natural coral implanted into bony tissue is gradually resorbed and replaced by newly formed bone [61].

Coral has been used clinically with good results in spinal fusion or to fill periodontal defects [62–64]. The exoskeleton of coral contained a high proportion of calcium carbonate, and this has been shown to be biocompatible, bioactive, osteoconductive and biodegradable at variable rates depending on the exoskeleton porosity, the implantation site, and the species [55, 65]. Natural coral, submitted to rigorous protocols of preparation and purification can be used as a replacement for bone grafts in both orthopaedic surgery and maxilla-cranial-facial surgery.

Shell

There are numerous shells in nature, and some of them have been evaluated for possible biomedical applications. Barnacle shells are very complex and strong composite bioceramic composed of different structural units which consist of calcite microcrystals of very uniform size. These are organized in a massive microstructure toward the external shell and in mineral layers separated by organic sheets toward the internal shell. It was observed that in contrast to nacre, barnacle shell contains calcite microcrystals (instead of aragonite microcrystals and has considerable porosity when compared with nacre which is a very dense material. The different composition and microstructural characteristics should affect dissolution behaviour in the body fluids (calcite being less soluble than aragonite)-and porosity could favour binding of this material with bone [66]. An investigation showed that calcium phosphate-based bioceramics have been synthesized by using eggshell-derived raw materials and phosphoric acid at different mixing ratios. The mineral composition of cockle shell is almost similar to that of coral and it was suggested that cockle shell has the potential to be used as a material for orthopaedic applications [53, 65].

The shell formation process in molluscs is a promising model for the development of bioinspired ceramics for a wide variety of applications in fields as varied as adaptive surface coatings, corrosion inhibition, hybrid composite materials, and more. The hybrid nature of the mollusc shell in terms of both mineral and organic components makes it an ideal bioceramic materials candidate: it is porous, it is a result of sophisticated structure, and exhibits exceptional flexural, fracture strength, and toughness [66].

Cowrie shells of West African origin are currently being evaluated for possible applications in orthopaedics as well as dental materials. The preliminary results have shown that these shells have great potential as biomedical materials.

Soil

It can be deduced from Table 1 that most of the calcium phosphates are inorganic, and they do exist as minerals. It was based on this that one of the authors believed that there is the possibility of obtaining these materials from nature The investigation of some naturally occurring clays show that they contain some of these minerals, and also the fact that some natural practitioners employed a special type of clay in bone fracture and spinal management. Currently, work is on-going on the possibilities of using anthill as a biomedical material.

BIOPOLYMER

Polymers play a central role both in the natural world and in modern industrial economies. Some natural polymers such as nucleic acids and proteins carry and manipulate essential biological information, while other polymers like polysaccharides—that is nature's family of sugars—provide fuel for cell activity and serve as structural elements in living systems [67]. The advantages of synthetic polymers include predictable properties, batch-to-batch uniformity, and they can be tailored easily but they are too expensive. The growing reliance on synthetic polymers has also raised a number of environmental and human health concerns. It was concluded that the focus should now be on natural polymers which are inherently biodegradable and can be promising candidates to meet different requirements [67, 68]. Emerging applications of biopolymers range from packaging, industrial chemicals, medical implant devices, to computer storage media. In this paper, the utilization of natural polymers biomedical materials will be considered. The major problems yet to be overcome with natural starting materials are their propensity for calcification and eventual biodeterioration [69]. Some of the disadvantages and advantages associated with natural biopolymers are as listed in Table 2. A study stated that polymers from natural sources are particularly useful as biomaterials; given their similarity to the extracellular matrix (ECM) other polymers in the human body [70]. Natural polymers include both ECM proteins derivatives (e.g., collagen) and some materials derived from plants and seaweed among others [71]. Biopolymers have wide applications as implantable biomaterials, controlled disease carriers or scaffold for tissue engineering.

Table 2: The advantages and disadvantages of natural polymers [72]

S/N	Advantages	Disadvantages
1	No problem with toxicity or foreign body response	A major issue is immunological reaction. Body's immune system recognizes foreign material and tries to destroy it
2	Can function biologically at molecular level, not just macroscopic level	High natural variability
3	If desired, natural degradation can occur in the body via natural enzymes. Can also add cross-links to make less degradable	Structurally more complex than traditional materials. Technological manipulation is more elaborate

Throughout history, humans have relied extensively on biological materials like wool, leather, silk, and cellulose, and these materials are natural polymers. Polymers are substances composed of repeating structural units that are linked together to form long chains [67].

Economics and Applications

Biopolymers are a diverse and versatile class of materials that have potential applications in virtually all sectors of the economy; typical examples are adhesives, absorbents, lubricants, soil conditioners, cosmetic, drug delivery vehicles, textiles, high-strength materials, and even computational switching devices [67]. The structure and some functions of natural polymers are as shown in Figure 1 [72]. It is reported that commercially available biopolymers are quite expensive but their application in specialized fields such as medical materials justified the relatively high costs. Also, the commercialization difficulties facing biopolymers in many ways resemble the problems confronting other emerging technologies such as photovoltaic cells and fuel cells. In this study, the utilization of biopolymers from natural products in the area of medicine is reviewed.

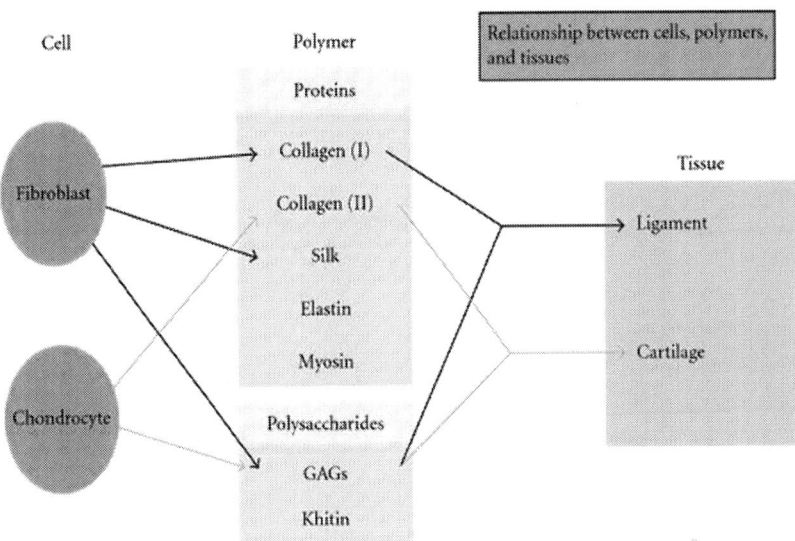

Figure 1: Natural polymers: structure and function.

Demand

It was forecast that natural polymer demand is expected to grow by 5.9% annually to $3.6 billion in 2010, reaching £1.7 billion. It was also stated that the threats to further growth include mature applications and variable supplies for products due to climatic and political uncertainties [73]. Polymers are regarded as a class of giant molecules consisting of discrete building blocks linked together to form long chains. Simple building blocks are called monomers, while the more complicated building blocks are sometimes referred to as "repeat units."

Generally, biopolymers fall into two principal categories:
- Polymers that are produced by biological systems such as microorganisms, plants and animals, in short natural products;
- Polymers that are synthesized chemically but are derived from biological starting materials such as aminoacids, sugars, and natural fats or oils. This can sometimes be referred to as synthetic natural product based polymers.

In this study, the emphasis is only on the first category of biopolymers. Exhaustive classifications are dealt with and their potential applications. Many types of polymers are widely used in biomedical devices that include orthopaedic, dental, soft tissue, and cardiovascular implants. Polymers represent the largest class of biomaterials, and they may be derived from natural sources, or from synthetic organic processes. The wide variety of natural polymers relevant to the field of biomaterials includes plant materials such as cellulose, sodium alginate, and natural rubber; animal materials like tissue-based heart valves and sutures, collagen, glycosaminoglycans, heparin and hyaluronic acid; and other natural materials, deoxyribonucleic acid (DNA), the genetic material of all living creatures [74]. Table 3 lists the various types of naturally occurring biopolymers and defines them on the basis of the chemical structure of their monomeric units and indicates the functions that these polymers serve in living organisms.

Table 3: Biopolymers found in nature and their functions [67]

Polymer	Monomer	Function(s)
Nuclei acids (DNA and RNA)	Nucleotides	Carriers of genetic information, universally recognized in all organisms
Proteins	-aminoacids	Biological catalysis (enzymes), growth factors, receptors, structural materials (wool, leather, silk, hair, and connective tissue); hormones (insulin); toxins; antibodies
Polysaccharides (Carbohydrates)	Sugars	Structural materials in plants and some higher organisms (cellulose and chitin); energy storage materials (starch and glycogen); molecular recognition (blood types); bacterial secretions
Polyhydroxylalkanates	Fatty acids	Microbial energy reserve materials
Polyphenols	Phenols	Structural materials in plants (lignin); soil structure (peat and humics); plant defense mechanisms (tannins)
Polyphosphates	Phosphates	Inorganic energy storage materials
Polysulfates	Sulfates	Inorganic energy storage materials

Natural polymers were viewed to offer the following properties: they are often identical, very similar to macromolecular substances which the biological environment is prepared to recognize and deal with metabolically. As a result, the problems of toxicity and simulation of chronic inflammatory reaction, which provoked by many synthetic polymers are suppressed. Furthermore, there is capability for designing those biopolymers at molecular rather than the macroscopic level, equal biopolymers are degradable. However biopolymers suffer from immunogenicity, elaborate technological manipulation, decomposition or modification and natural variability in structure from the sources. These divergent factors have produced some unique biomedical materials with outstanding performance. Table 2 shows the general properties of certain natural polymers [75]. A comprehensive overview of the family of natural polymers is listed in Table 4. In this study the broad classification, presented in Table 5, will be adopted for biopolymers and they are grouped into proteins, polysaccharides, and polynucleotides. Polynucleotides are used for purposes such as nanostructural materials and assembling as well as for their electrical properties, but this is are not currently of interest in the field of biomaterials. Hence, this paper will be limited to proteins and polysaccharides.

Table 4: General properties of certain natural polymers [67]

Polymer	Incidence	Physiological Functions
(A) Protein		
Silk	Synthesized by anthropods	Protective cocoon
Keratin	Hair	Thermal insulation
Collagen	Connective tissues (tendon, skin)	Mechanical support
Gelatin	Partly amorphous collagen	Industrial product
Fibrinogen	Blood	Blood clotting
Elastin	Neck ligament	Mechanical support
Actin	Muscles	Contraction motility
Myosin	Muscles	Contraction motility
(B) Polysaccharides		
Cellulose	Plants	Mechanical support
Amylose	Plants	Energy reservoir
Dextran	Synthesized by bacteria	Matrix for growth of organism
Chitin	Insects and Crustaceans	Provides shape and form
Glycosaminoglycans	Connective tissues	Contribution to mechanical support
(C) Polynucleotides		

Deoxyribonucleic acids (DNA)	Cell nucleus	Direct protein biosynthesis
Ribonuclei acids (DNA)	Cell nucleus	Direct protein biosynthesis

Table 5: A Snapshot of the biopolymer family

Polyesters	Polysaccharides (fungal)	Lipids/surfactants
Polyhydroxyalkanoates	Pullulan	Acetoglycerides, waxes
Polylactic acid	Elsinan	Surfactants, emulsan
	Yeast glucans	
Proteins		Polysaccharides (bacterial)
Silks	Polysaccharides (plant/algal)	Xanthan
Collagen/gelatine	Starch (amylose/amylopectin)	Dextran
Elastin	Cellulose	Gellan
Resilin	Agar	Levan
Adhesives	Alginate	Curd lan
Polyaminoacids	Carrageenam	Polygalactosamine
Soy, zein, wheat glutein,	Pectin	Cellulose (bacterial)
casein, and serum albumin	Konjac	
	Various gums (guar)	Specialty Polymers
Polyphenols		Shellac
Lignin	Polysaccharides (animal)	Poly- -glutamic acid
Tannin	Chitin/chitosan	Natural rubber
Humic acid	Hyaluronic acid	Synthetic polymers from natural fats and oils (nylon from castor oils)

*Adapted from Herdman (1993) from p20.

Proteins

Proteins are polymers that are composed of amino acids. The specific amino acids used and the sequence of amino acids in a protein polymer chain are determined by the corresponding deoxyribonucleic acid (DNA) template. Proteins are also referred to as polypeptides, these are complex copolymers composed of up to 20 different amino acids building blocks. There are virtually a limitless number of proteins that can be formed from these 20 monomers. Proteins can contain a few hundred amino acid units or thousands of units. Protein polymers are unique in that the sequence of the monomers in the polymer chain is predetermined by the template-specific reamer of the polymerization process [14, 67].

Many proteins are of commercial interest because of their catalytic (enzymatic) or pharmaceutical properties. Nature has provided a vast array of proteins whose principal function is to form structural materials in living organisms. Some of the more familiar protein materials include wool, leather, silk, and gelatine. Also, elastomeric proteins occur in a wide range of biological systems where they have evolved to fulfil precise biological roles. The best known include proteins in vertebrate muscles and connective tissues, such as titin, elastin, and fibrillin, and spider silks. While some other examples are byssus and abductin from bivalve, molluscs, resilin from anthropods, and gluten from wheat [67, 76].

Gelatin

This is obtained through a controlled denature of the fibrous insoluble protein, collagen which is the major component of skin, bone, and connective tissue. It is characterized by having no antigenicity in comparison to its precursor. Gelatin is widely used as scaffold for tissue engineering and also has been frequently used in medicine as a wound dressing, and as an adhesive, absorbent pad for surgical use [77, 78].

Collagen

Collagen is an example of natural material which exists mostly in fibril form; it has a characteristic triple-helix structure and is the most prevalent protein in the animal world. It is the most abundant of all proteins found in mammals typically accounting for more than 30 percent of body protein. It forms a significant component of connective tissue such as bone, tendons, ligaments, and skin. There are at least 10 different types of collagen in the body. The arrangement of collagen fibres is arranged parallel to one another to give a structure with the tensile strength of a light steel wire. In skin, where strength and flexibility are required, collagen fibres are randomly oriented and woven together like felt [16, 67]. Collagen is a protein that acts as a structural support in a wide range of tissues including skin, bone, tendons, ligaments, cartilage, blood vessels, and nerves. It is usually implanted in a sponge form that does not have significant mechanical strength or stiffness. It has shown good promise as a scaffold for neotissue growth and is commercially available as a product for wound healing. Injectable collagen is widely used for the augmentation or buildup of dermal tissue for cosmetic reasons. In particular, collagen and its degradation products are often used for the attraction of fibroblasts in vivo during wound repair, fracture healing, and embryogenesis [16, 78, and 79].

Silk

Silk is natural fibrous protein which is spun by Lepidoptera larvae such as silkworms, spiders, scorpions, mites, and flies [80]. Several attempts to produce silk by bacteria, yeast and plants by inserting genes in them gave insoluble silk proteins, which slumped inside cells [13].

It is well known that silk fibres are composed of at least two main proteins: sericin and fibroin. Silk fibroin shows excellent physical and chemical properties. Silk without sericin showed higher stability than others. Also, silk fibroin can be prepared in various forms; gel, powder, film, matrix, or fibre depending on the applications [80, 81]. Silk has always been a material of great fascination. Spiders can process silk protein into a material that has a tensile strength 16 times greater than that of nylon and a very high degree of elasticity. Silkworm silk is about

2 or 3 times greater than that of nylon. Silk also possesses the ability to super contract especially when they are put into liquid [67]. The breaking stress and strain for different silks are presented in Table 6 [13].

Table 6: Stress and strain of different types of silk

Silk type	Stress (Gpa)	Stress (Gpa)
Bombyx mori cocoon silk	1.1	0.24
Nephila claripe MAS	1.75	0.15
Nephila maculate dragline	1.1	0.46
Arameus serratus framesilk	0.81	0.24
Araneus diadematus radial silk	1.2	0.40
A. serratus viscid silk	1.0	2.00
A. diadematus viscid silk	1.4	4.76

*Adapted from Ivanova (2005) [13].

It is observed that the most interesting silk for engineering is viscid silk of Adameus Diamantus since it has greater stress and strain at breaking point. The strength and toughness of silk is known to be remarkable, and some of the mechanical properties of biodegradable materials are apresented in Table 7 [11].

Table 7: Mechanical properties of biodegradable materials

Source of biomaterial	UTS (MPa)	Modulus (GPa)	Strain (%) at breakage
Bombyx mori silk (with sericin)	500	5–12	19
Bombyx mori silk (without sericin)	610–690	15–17	4–16
Bombyx mori silk	740	10	20
Collagen	0.9–7.4	0.0018–0.046	24–68
Cross-linked collagen	47–72	0.4–0.8	12–16
Polylactic acid	28–50	1.2–3.0	2–6

*Adapted from Cao and Wang (2009) [11].

Generally, silk has been investigated for use as biomedical resource due to its unique properties which include nontoxicity, biocompatibility, and biodegradability. It must be noted that it is quite hard and expensive to produce silk from these natural sources in large quantities. Silk has been used as a biomaterial in various forms such as films, membranes, gels, sponges, powders, and scaffolds. They have been commercially applied as scaffolds, vascular prostheses, structural implants, nets, and sutures among others [30, 31, and 67].

Keratins

The structural integrity and solubility of keratin as well as its natural biocompatibility, controllable biodegradability, and bioactivity makes it an ideal material medical polymer. A proposal was made that keratin extracts from hair and wool could be used as platform technology to make a new family of biomaterials used for biomedical applications such as wound healing and bone regeneration, scaffold for tissue engineering, and coatings for medical devices. This proposal was a challenge to the long-standing notion that these animal-derived proteins would not be compatible with human biological systems. It was proved that the carefully extracted keratin molecule did not elicit an adverse biologic response [82]. These are a class of biomaterials that can be derived by extraction of proteins from human hair. They have haemostatic characteristics, and it was hypothesized that a keratin hydrogel having the ability to absorb fluid and bind cells may be an effective haemostat [83]. Also the usage of keratins to mediate a robust nerve regeneration response in part through activation of Schwann cells was reported [84]. They also suggested that keratins derived from human hair are neuroinductive and can facilitate an outcome comparable to autograft in a nerve injury model.

Elastin

The best known and most widely distributed protein elastor is elastin. It is responsible for the elasticity of the aorta and skin of mammal, and is also present in the ligamentum nuchae which is involved in raising the heads of grazing hoofed animals. Elastic fibres are elastic, load-bearing protein polymers found in connective tissue such as ligaments. This rubber-like material responds to changes in temperature and is able to convert chemical energy into mechanical energy. As a consequence, the material could be used as a replacement for ligament tissue, blood vessels, or any other tissues requiring the contractile properties of elastin [67, 76].

Natural Rubber Latex

It was recently claimed that the natural rubber latex, extracted from Hervia brasilliens, is a strong candidate for use as a biomaterial. This biologically compatible material can be used in contact or inside the human body, and it has been shown that natural rubber latex performs a biological action that accelerates the healing process, being a powerful stimulator of cicatrisation. It is well known that dried latex, in the rubber form, presents high resistance; high elasticity; and is an easy-shaping material, which increases its appeal for use in the fabrication of vascular prostheses [85, 86].

Polysaccharides

Polysaccharides are inexpensive, natural biopolymers which are widely used as raw materials in several industries and being a natural compound, most polysaccharides are easily biodegradable. 75% of all organic material on earth is present in the form of polysaccharides [87, 88]. Polysaccharides are polymers or macromolecules composed of simple sugars, and they have two principal functions. Some such as cellulose serve as structural materials in living systems. Polysaccharides can be utilized as materials in medicine which include wound management, tissue engineering, drug delivery systems, and haemostatic devices among others. It can be classified as listed in Table 8.

Table 8: Types of polysaccharides

Types	Examples
Bacterial	Xanthan, dextran, gellan, and cellulose
Fungal	Pullulan, elsinan, yeast, and glucan
Plant	Starch, alginate, and tannin
Animal	Chitin, chitosan, hyaluronic acid, lignin, and tannin

Starch

Starch is one of the most abundant and cheap polysaccharides. Natural starch occurs in a granular form, and it is a principal carbohydrate storage product of plants. It is found in cereal and tuber plants such as maize and potatoes, respectively [67, 87, 89]. The content of amylose and amylopectin in starch varies and largely depends on the starch source. Most often, starch consists of about 30% amylose [a linear α-(1–4) glucan] and 70% amylopectin (dendritically branched version) [89–92]. The structure of starch is shown in Figure 2, and it is adapted from Nair and Laurencin [93]. The ratio of amylose and amylopectin in the starch may affect the starch behaviour during processing and the properties of the end product. As the amylose content increases, it will also increases the crystallinity of starch based products and this resulted in texture firming. It must be emphasized that starch is an established and widely used biodegradable polymer. Also starch possesses some properties which enable it to be compounded with other biopolymers with resultant improved products. Starch is insoluble in cold water, but it is very hygroscopic and binds water reversibly. It is renewable, biodegradable, inexpensive, and as such can play an important role in the medical field. They are studied in several biomedical applications ranging from bone replacement implants, bone cements to drug delivery systems and tissue scaffolds [94, 95].

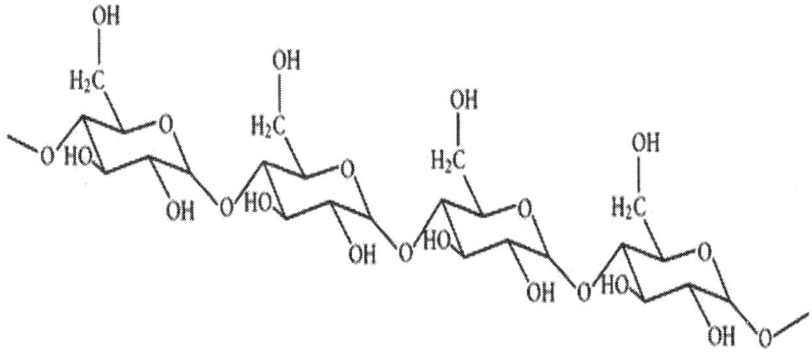

Figure 2: Structure of starch.

Chitin and Chitosan

Chitin is a nitrogen-containing polysaccharide, related chemically to cellulose, and it is insoluble in most solvents. Controlled deacetylation is used to produce chitosan which is a derivative with approximately 50% free amine. Chitins are regarded as the second most abundant natural polymer after cellulose and there are three possible sources of chitin as raw materials. They are isolation from traditional shellfish sources such as crabs and shrimps; harvesting of fungal mycelia from bioreactor processes, a typical example is mushroom; synthesizing from monomeric/dimeric units using chemical and/or enzymatic strategies. However in this paper, the third source of obtaining chitin will be neglected because it involves chemical synthesis. Recently, studies have shifted from the traditional shellfish to exciting and novel shell materials like eggshell and cowry shell. This may be attributed to the environmental pollution and the concern for the amount of residual protein [96–98].

Chitosan is currently obtained by the deacetylation of chitin-9poly-β-(1–4)-N-acetyl-D-glucosamine). Chitosan (1, 4 linked 2-amino-2-deoxy-β-D-glucan) comprises of glucosamine and N-acetylglucosamine. The latter is the water soluble chitin called chitosan [99], which is a moiety of glycosaminoglucans (GAGs). The structure of chitosan and chitin is shown in Figure 3 [93]. Meanwhile, chitin and chitosan are regarded as natural resources waiting for a market and sourced from the canning industry especially the crustaceans [100]. They stated that there are four chronological steps by which chitosan can be processed from crustacean shells. The steps are as shown in Figure 4, and they are, namely, deproteination, demineralization, decolouration, and deacetylation. Chitin and chitosan are biologically stemmed aminopolysaccharides which exhibit multiple bioactivities, for example, low toxicity, biocompatibility, biodegradable, antimicrobial, and wound-healing properties. Moreover, chitosan elicits minimal foreign body reaction. Chitosan favours both soft and hard tissue regeneration [80, 101].

Natural Products: A Minefield of Biomaterials 263

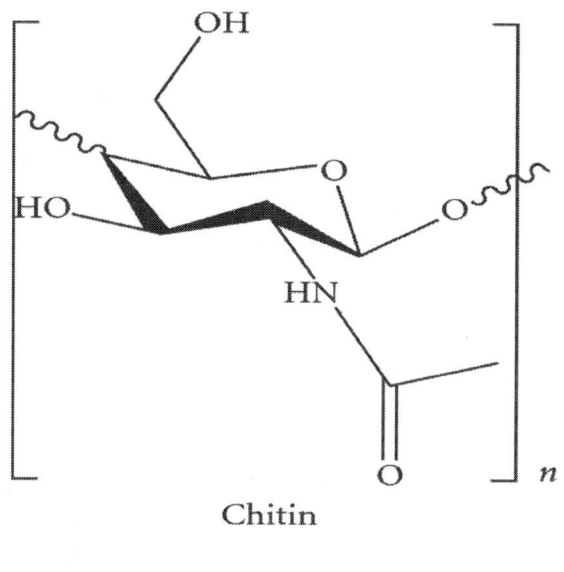

(a)

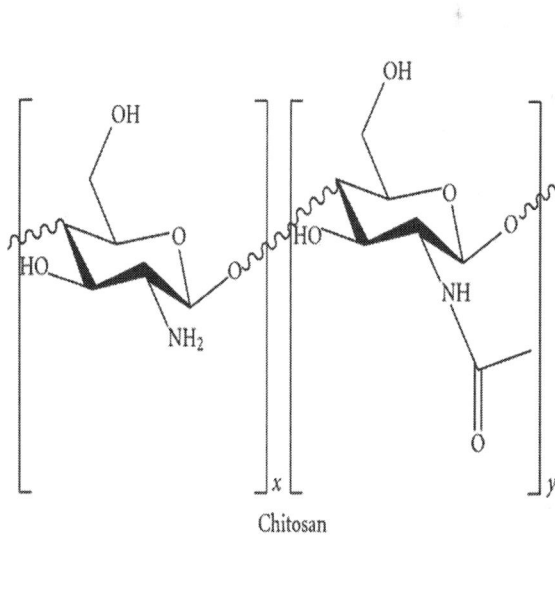

(b)

Figure 3: Structure of chitin (a) and structure of chitosan (b).

Crustacean shells ⟶ size reduction ⟶ protein separation ⟶ (NaOH) ⟶ washing demineralization (HCl) ⟶ washing and dewatering ⟶ decolouration ⟶ chitin ⟶ deacetylation (NaOH) ⟶ washing and dewatering ⟶ chitosan

Figure 4: The processing stages in producing chitosan from crustacean shells.

Chitins have various biofunctionalities including antithrombogenic, haemostatic, immunity enhancing, and wound healing. Research has shown that chitin and chitosan are nontoxic and nonallergenic, so the body does not reject these compounds as foreign invaders. Biocompatibility, biodegradability, and adsorption properties of chitin and its derivatives are much higher than synthetically substituted cellulose. The native chitin molecule has strong inter- and intra-molecular hydrogen bonds with partial N-deacetylation [102]. Chitin is a biocompatible and biodegradable polymer which demonstrates bacteriostatic and analgesic effects, which in addition, shorten the time of wound healing and the rebuilding of connective tissue. The only disadvantage of natural chitin application is its insolubility in the common solvents generally available, and the enormous related difficulties connected with chitin processing [103].

The chitin family of polymers is being widely used in various applications such as medicine, manufacturing, agriculture, and waste treatment. In the biomedical area, chitosan and its derivatives have been successfully used in wound dressings, drug delivery systems, and as materials for tissue engineering. They have been reported to be a promising candidate as a scaffold material for engineered human tissue such as skin, cartilage, and bone due to its biocompatibility, and resorbability. Chitosan is used as an intraocular lense material because of its oxygen permeability and it has also been found to expedite blood clotting. It is also being evaluated for use in the bioremediation of toxic phenolic compound [67, 104, and 105].

Hyaluronic Acid

Hyaluronic acid (HA) is a natural polysaccharide polymer belonging to the same class of compounds as starch and cellulose. The substance is found naturally in the extracellular matrix of skin, cockscomb, cartilage, vitreous humor, and other body tissue and plays a role in the movement and proliferation of cells. It also occurs as an extracellular polysaccharide in a variety of bacteria [67, 106]. HA was discovered in 1934, and it is a long unbranched polysaccharide chain, composed of repeating twin sugar units. Due to the high density of negative charges along the polymer chain, HA is very hydrophilic and adopts highly extended random coil conformations. It is extremely flexible, has a high viscosity, and structure is shown in Figure 5[67, 93]. HA consists of 2-acetamide-2-deoxy-α-D-glucose and β-D-gluconic acid residues linked by alternate (1,3) and (1,4) glycoside bonding and has the high capacity of lubrication, water sorption, and water retention, and influences several cellular functions such as migration, adhesion, and proliferation. HA is water-soluble, and must be cross-linked or otherwise modified to form a scaffold [71, 78].

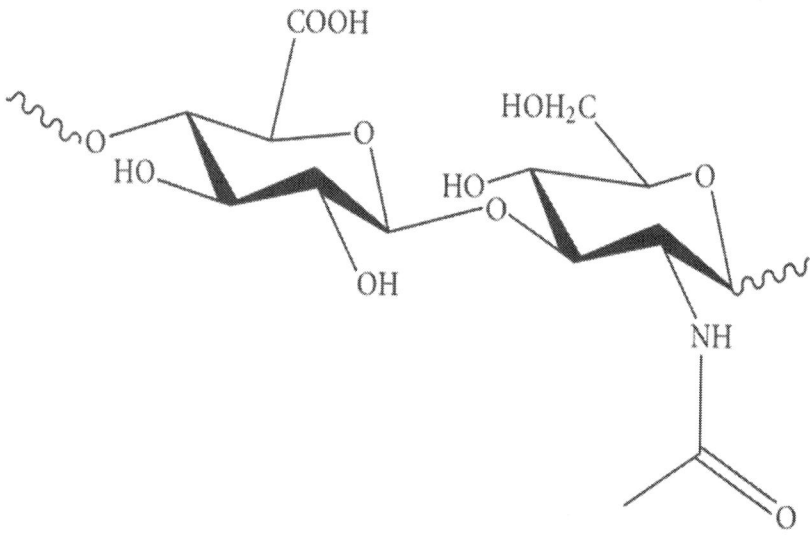

Figure 5: Structure of hyaluronic acid.

It is an extremely attractive polymer material because it is a natural product that degrades into simple sugars. Recent biomedical applications of hyaluronic acid include scaffolds for wound healing and tissue engineering, as well as ophthalmic surgery, arthritis treatment, and as a component in implant materials. It is also used as a lubricating fluid in joints and serves as a regulator in the lymphatic system. Based on its in vivo functions, HA has been adapted to commercial compounds for treating a number of tissue-related conditions, including osteoarthritis of the joints, and facial wrinkles and folds of the skin [67, 78, and 107].

Pullulan

Pullulan is water-soluble polysaccharide produced outside the cell by several species of yeast, most notably Aureobasidium Pullulans. It is a linear polymer made up of monomers that contain three glucose sugars linked together [67, 108]. It is claimed that pullulan is a nonionic exopolysaccharide of fungal origin and is currently exploited in the food industry due to its many unique characteristics [109]. It is nontoxic, nonimmunogenic, nonmutagenic, and noncarcinogenic; hence there are attempts to explore pullulan for various biomedicals [110].

Pullulan acts as a plasma extender without undesirable side effects. After metabolic turnover, it is completely excreted, and it was reported that pullulan to be used as plasma blood expander should have a molecular weight of about 60 kDa. Pullulan is biodegradable, impermeable to oxygen, nonhygroscopic, and nonreducing. Recently, the role of polysaccacharides in developing controlled drug delivery systems has increased significantly, and pullulan is gaining lot of attraction for this application [111]. Tissue engineering requires scaffolds or artificial extracellular matrix (ECM) that can accommodate cells and regulate their growth leading to three-dimensional tissue regeneration. It was concluded that heparin-conjugated pullulan material can thus be used for the proliferation of vascular endothelial cells and to inhibit the proliferation of smooth muscle cells (SMCs) [108]. Pullulan has very important applications in surface modification of polymeric materials so that they can be made more blood compatible and bioinert. Pullulan being adhesive is also tried for wound dressing applications. The thermal stability and elastic properties of pullulan allow it to be utilized in many different ways [109, 112, and 113].

BIOCOMPOSITE

Naturally, animals and plants synthesize biocomposites with high strength consisting of fibrous biopolymers. A classic example is cellulose which consists of whisker like microfibrils that are synthesized and deposited in a definite manner which imposes high strength. Living tissues are composites themselves with a number of levels of hierarchy. Reinforcement of polymer with nanosized particles or fibres is a promising technique that is capable of yielding materials with enhanced performance but without involvement of expensive synthesis procedures [103]. Bone is regarded as a biocomposite which comprises of mainly nanohydroxyapatite crystallite orienting along collagen fibres [77].

There is no really adequate definition of a composite material; however, there are three criteria that have to be satisfied before a material can be said to be a composite. Firly, the constituents have to be present in reasonable proportions and secondly, the constituent phases must have different properties and as such ultimately leading to properties that are superior and possibly unique in some specific respects, to the properties of the individual components. Lastly, the processing technique must be (in such a way that) able to be carried out under controlled conditions. There are several means by which composites can be classified and in this paper, we are interested in natural composite materials. Some typical examples of this class are wood, bone, muscle, and other tissues. Bone is a composite material composed primarily of organic fibres, small inorganic crystals, water, and fats. The proportions of these components will vary with the type of bone, animal species, and age, but typically about 35% of the dry, fat-free weight of bone is the organic fibre, collagen. Collagen is a fibrous protein and located around the outside of the collagen fibres are small rod-like crystals of hydroxyapatite with dimensions of the order of 5 × 5 × 50 nm. Therefore at this microscopic level we have a hydroxyapatite-reinforced collagen composite and this may be considered to be the basic "building block" of bone [114, 115].

It must be realised that to classify composite within the context of this work is severely restricted. This is due to the fact that only combinations of biopolymers and bioceramics are possible. Thermodynamically, metals are stable in combined state as ores (i.e., oxides, sulphides, and carbonates), and they rarely exist in their pure form; hence only precious metals are available as natural materials for biomaterials application. This paper will consider combination of biopolymer with biopolymer, biopolymer with bioceramics; there have been little or no reports on the application of two bioceramics as biocomposites.

Biopolyme-Biopolymer

Biopolymers occur abundantly in nature, and they have several applications as materials in medicine. Chitosan is regarded as a biodegradable and nontoxic hydrophilic polysaccharide with excellent mucoadhesive and permeation-enhancing properties. Alginate is the name given to a family of linear polysaccharides found in brown algae and is composed of guluronic and mannuronic units [106]. Meanwhile, there has been increasing interest in the study of alginate-chitosan microparticles as carriers for controlled release of proteins and drugs due to its biocompatible, biodegradable, and mucoadhesive properties. It was shown that the stability of alginate-chitosan capsules depended strongly on the amount of chitosan bound to the capsules [116]. The superporous hydrogel foams containing chitosan and gelatine have a lot of interesting food and medical applications. The superporous structure and mechanical properties are especially proper for scaffold preparation [105].

Chitosan/gelatine hydrogels are being developed for tracheal epithelia. A company in Italy is reported to have developed a series of modified hyaluronate esters by adding hydrophobic moieties to the carboxyl groups so as to control degradation, and it is marked as tissue engineering polymers which are being applied to bone growth and cartilage [71]. An investigation concluded that the artificial dermal skin is composed of gelatine and polysaccharides such that hydraluronic acid and β-glucan will be useful to promote wound healing [78]. While a combination of flexible protein and rigid polysaccharide results in a series of biomimetic chitosan/gelatine based biomaterials covering surface modifier and nonviral vector for gene therapy. They hold promising prospect in tissue engineering [77].

Cellulose-based hydrogels are advantageous over cellulose sponges and fabrics as their bulk chemistry can be easily modified. Independent investigations have reported the synthesis of novel biomimetic hydrogels, based on crosslinking cellulose derivatives with hyaluronic acid [117, 118]. Although mainly proposed as postoperative adhesion barriers, such hydrogels also show potential as scaffolds for regenerative medicine, with a tunable degradation rate. Indeed, the presence of hyaluronic in the cellulose network provides enzyme-sensitive degradation sites, whose density in the bulk of the hydrogel can be easily controlled [119].

The blending of polymers with starch under controlled conditions leads to copolymerization that in turn results in high-molecular polymers with thermoplastic properties. Though, the mixing or blending needs special machinery, such as an extruder, the products (polymer blend) can be handled as easily as conventional plastic resin [94]. Although starch such as that of cassava is used as a polymer, other polymers are frequently used in the blend. Polycaprolactone blend is the most commonly used polymer/starch blend because of its low-melting temperature (Tm) and high susceptibility to amylase and lipase hydrolyses [120,121]. It must be noted that starch as polymer has been marketed under various names, and most patents describe a generic starch, this includes cassava [100].

Natural polymers such as chitosan and cellulose, and their derivatives are inherently biodegradable, and exhibit unique properties. Starch and chitosan are abundant naturally occurring polysaccharides. Both of them are cheap, renewable, non-toxic and biodegradable [122]. Bourdoom and Chinnan (2008) concluded that the starch/chitosan blend exhibits good mechanical properties, while water barrier properties, and miscibility of biodegradable blend films are affected by the ratio of starch and chitosan.

One problem associated with starch-based blend is that starch and many polymers are nonmiscible which leads to the mechanical properties of the starch/polymer blends generally becoming poor. Thus, chemical strategies are taken into consideration in order to improve the properties [123]. Starch-based biodegradable polymers have some advantages for use as medical materials such as good biocompatibility, biodegradable, and its degradation products are non-toxic, proper mechanical properties and degradation as requirement. Starch-based biodegradable polymers have been widely investigated in bone tissue engineering.

Starch-based biodegradable bone cements can provide immediate structural support and degrade from the site of application. In addition, starch-based biodegradable polymers can also be used as bone scaffold and when they are in the form of microsphere or hydrogel they are suitable for drug delivery systems. There is no need for surgical removal of the device after drug depletion. The unique properties such as hydrophilicity, permeability, biocompatibility, are to some extent similar to soft biological systems, all these make starch-based hydrogel to be useful for various biomedical applications [124–129]. Biodegradable polymers such as poly (lactic acid), poly (glycolic acid), and their respective copolymers have been used in several drug delivery systems. However, few attempts have been made to use starch-based polymers in these types of applications; despite being well known that they are biodegradable materials; they have been proposed in several works to be used as biomaterials [130–133].

Biopolymer-Bioceramics

A composite made of bioceramic-polymer composite was introduced, and it was based on the concept that cortical bone itself comprises an organic matrix reinforced with a mineral component. The material they developed has been used as an orbit implant for orbital floor fractures and volume augmentation; also it is now being used in middle ear implants, commercialized under the trade name HAPEX [134, 135].

Polymers such as alginates showing low toxicity and high biocompatibility and hydroxyapatite have been reported with osteoinductive or osteoconductive potential. Composites binding these properties could be a good choice for the development of new materials for medical application. This composite is considered to be biocompatible and partially resorbable [136].

The association of bioceramics and fibrin sealants may develop the clinical applications of bone substitutes. The physical, chemical, and biological properties of both bioceramics and fibrin glue may be cumulated for preparing advance bone substitutes. The ideal bone substitute should be biocomposite, biodegradable at the expense of bone growth and mouldable, with sufficient mechanical properties to fill and restore bone defects [137]. These biological properties can be tailored for both bioceramics and fibrin glue by changing the composition; porosity and network cross-linking. The positive effects of the combination of fibrin sealant with ceramic biomaterials have essentially been observed in clinical studies. Fibrin sealants improve the surgical handling of biomaterials and widen their field of application in bone surgery. Furthermore, the future development of this composite may be in combination with bone growth factors [138].

One of the naturally occurring biocomposites that has been deployed as biomedical materials is nacre. It is secreted only by the molluscan classes, Gastropoda, Bivalvia, Cephalopoda, and to a minor extent, Tryblidiida. Nacre has a lamellar structure consisting of alternating tablets of aragonite and organic interlamellar membranes which have a core of β-chitin surrounded by acidic proteins [139]. It is now clear that the sequence of nacre formula involves the secretion of interlamellar membranes separated by a liquid rich in silk fibroin such that subsequently the liquid is replaced with mineral. This pattern is the same for the bivalves and gastropods, so too for the other nacre-secreting molluscs, although this is yet to be determined. There are, however, structural differences between bivalve and gastropod nacre [140].

Nacre is by far the most intensively studied nonhuman organomineral biocomposite. It has a high proportion, approximately 5% of organic matter (proteins and polysaccharides), the mineral fraction being exclusively in the form of aragonite. The aragonites work of fracture is estimated to be as high as 3000 times that of inorganic aragonite but this figure is considered to be lower in reality. Nacre has found possible applications in the biomedical industry due to their superior biomechanical properties [140]. Nacre is able to form a tight bond of bone without soft and fibrous tissue formation. It is gradually and centripetally dissolved but not resorbed by cells because it is not porous [141, 142].

APPLICATION OF NATURAL PRODUCTS IN BIOMATERIALS

There are various uses of natural products in materials for medical applications, some of which are bioelectrodes, dental implantation, orthopaedic applications, adhesive and sealant, ophthalmological applications, intraocular lens implants, burn dressings, and skin substitutes among others. Table 9 lists some of the natural biomaterials and their areas of application [10]. There are three principal market segments for natural biomaterials: wound management products, drug delivery systems (DDSs), and tissue engineering. In this paper, all these three will be discussed and another important area of application will be highlighted. Biotextiles have utilized natural product-based proactive agents such as chitosan, natural dyes, neem extract, and other herbal products for antimicrobial finishing of textile substrates [143]. Biotextiles also provide a good platform to demonstrate the limitations and challenges of natural products as materials in medicine.

Table 9: Some applications of natural biomaterials

Application	Natural biomaterial
Artificial heart valves	Bovine pericardium, Intact porcine aortic valves
Hernia repair devices	Porcine small intestinal submucosa, porcine urinary bladder mucosa, Porcine dermal grafts
Sutures	Catgut (porcine or bovine intestinal wall) and porcine dermal grafts
Skin repair/ wound care	Dermal allograft, porcine small intestinal submucosa, and porcine dermal grafts
Vascular prostheses	Bovine ureter, porcine small intestinal submucosa, and Ovine arteries
Urethral repair	Porcine bladder
Breast reconstruction	Dermal allograft
Ligament repair	Dermal allograft, porcine small intestinal submucosa, and fetal bovine skin
Spinal fusion/ bone healing	Bone allograft

Wound Management

This segment is regarded to grow at an astronomical rate, and it represents a sizeable portion of biomedical materials. Hyaluronic acid is an extremely attractive polymer material used in wound-healing preparations, aids tissue formation and repair, provides a protective matrix for reproductive cells, serves as a regulator in the lymphatic system, and acts as a lubricating fluid in joints. Zein, which is a major storage protein of corn, has been used widely as an adhesive, fibre, cosmetic powder, and ink [67]. Hide glue derived from gelatin has been in used for centuries. Lastly, mention must also be made of burn dressings methods like autografts, allografts from cadavers, and xenografts of which domestic swine is the most commonly used as a temporary wound closure material [14].

Drug Delivery System

The discovery of new drugs has been the major thrust in drug research, and although this trend will continue for a while, there is an increasing emphasis being placed on the development of novel drug delivery systems. The use of biopolymer materials for drug delivery can minimize tissue reaction and allow drugs to be administered in nonconvectional ways. The use of biopolymers in these formulations has thus far been restricted to a narrow set of applications [67]. Natural polysaccharides due to their outstanding merits have received more and more attention in the field of drug delivery systems. In particular, polysaccharides seem to be the most promising materials in the preparation of nanometric carriers. In nature, polysaccharides have various resources from algal origin (e.g., alginate), plant origin (e.g., pectin and guar gum), and microbial origin (e.g., chitosan and chondroitin). Among various polysaccharides, chitosan is the early one to be used to prepare nanoparticles [144, 145]. Also, the main applied form of albumin for biomaterials is pharmaceutical microspheres, which has been extensively investigated for drug targeting to various organs and tissues [146, 147].

Tissue Engineering

One key area of research gaining significant attention over the past several years is tissue engineering. This technology combines an engineered scaffold, or 3-dimensional structure with living cells. These scaffolds can be constructed of various materials and made into different shaped depending on the desired application [14]. Many naturally occurring scaffolds can be used as biomaterials for tissue engineering purposes, and one of the typical examples is hydroxyapatite which is obtained from coral or animal bone. Proteins are one of the important candidates for tissue engineering. They provide cell support for anchorage and adherence through cellular growth and development [148, 149]. Currently available proteins for application in tissue engineering include fibrin, collagen, zein, silk fibroin, keratin, casein, and albumin. A more speculative application is the use of biopolymers as scaffolding in the formation of new cartilage in the body. With the advantage of biocompatibility, biopolymers are likely to be used in many more novel orthopaedic applications [67].

Medical Fibres and Biotextiles

The application of fibres and biotextiles as components for implantable devices is widespread and covers all aspects of medicine and healthcare. They are used as drapes and protective apparel such as protective surgical gowns, operating room curtains, masks, and shoe covers. Textiles can also be used as topical and percutaneous applications in areas like bandages, wound coverings, and nappies. Lastly, three key applications of biotextiles in general surgery are sutures, haemostatic devices, and hernia repair meshes. Catgut is a natural collagen-based suture material obtained from ovine intestine, which is cross-linked and cut into narrow strips, and it is one of the first bioabsorbable fibres used in surgery. Silk and collagen are two natural fibres that have been widely used in medicine and multiple applications. Silk is inexpensive and is considered to be the gold standard in suture-handling characteristics. Collagen sutures are primarily used in microsurgery or ophthalmic surgery. Cotton was and still is commonly employed for bandages, surgical sponges, curtains, and surgical apparel, and in surgical gowns [14].

Antimicrobial textiles help to reduce effectively the ill effects associated with microbial growth on textile material. The use of natural products such as chitosan and natural dyes for antimicrobial textile materials has been reported. Plant products comprise the major segment of natural antimicrobial agents; sericin, natural dyes, tannins, and aloe vera among others. Aloe vera also possesses antifungal and antibacterial properties, which can be exploited for medical textile applications, such as wound dressing, suture, and bioactive textiles. The major challenges in application of natural products for textile application are that most of these biomaterials are complex mixtures of several compounds, and also the composition varies in different species of the same plant. The activity and composition also vary depending on their geographical location, age, and method of extraction. The availability of these products in bulk quantities, their extraction, isolation, and purification to get standardized products are other challenges in their applications [143].

CONCLUSIONS

Natural products have found many applications in the field of pharmaceuticals and medicine; healthcare and nutritional supplements; agrochemicals among others. The paper looks at the recent research in natural biomaterials towards applications in various biomedical fields. The purpose of this paper is to highlight information available on the various forms of natural products for biomaterials as well as to highlight the applications of natural biomaterials. The properties, demand, and economic importance of different natural biomaterials are discussed. Three major areas of intervention of natural biomaterials were discussed. It is advocated that future work should be focussed on the methods of reducing the major disadvantages like poor immunogenic response, variability, and the technological processing techniques. The poor immunogenic response is attributed to the presence of antigenic determinants which can be reduced by either chemical modification or standardized natural material sourcing. There is a need to develop a well-tested production and processing route for producing the biomaterials from the natural products as this will reduce the effects of variability among others.

In this paper, an attempt has been made to increase the understanding of the utilization of natural products in the biomedical field. This will attract the attention of specialists in the study of natural products and biomaterials.

REFERENCES

1. J. P. A. Nicolai and G. Rakhorst, Introduction: Biomaterials in Modern Medicine, 2008, http://www.worldscibooks.com/etextbooks/6562/6562.chap01.pdf.
2. G. Heness and B. Ben-Nissan, "Innovative bioceramics," Materials Forum, vol. 27, pp. 104–114, 2004.
3. D. F. Williams, Definition in Biomaterials, Elservier, Amsterdam, the Netherlands, 1987.
4. A. F. Von Recum and M. LaBerge, "Educational goals for biomaterials science and engineering: prospective view," Journal of Applied Biomaterials, vol. 6, no. 2, pp. 137–144, 1995.

5. L. Sedel, "Biomaterials: Medical Viewpoint," 2004, http://www.atsmp.whut.edu.cn/resources/pdf/4078.pdf.
6. J. P. Brown, M. Bide, and M. Phaneuf, "Cellular encapsulation into porous alginate fibers," Tech. Rep. M04-CL13, National Textile Center Research Briefs—Materials Competency, 2005.
7. L. M. Zimmerman and I Veith, Great Ideas in the History of Surgery, Norman Publishers, New York, NY, USA, 1st edition, 1993.
8. DSM, Corporate Communications, A Brief History of Biomedical Materials, 2009, http://www.dsm.com/en_US/cworld/public/medical downloads/.
9. R. H. Doremus, "Review—bioceramics," Journal of Materials Science, vol. 27, no. 2, pp. 285–297, 1992.
10. J. C. Coburn and A. Pandit, "Chapter 13: development of naturally-derived biomaterials and optimization of their biomechanical properties," in Topics in Tissue Engineering, supplement 13, pp. 1–23, 2007.
11. Y. Cao and B. Wang, "Biodegradation of silk biomaterials," International Journal of Molecular Sciences, vol. 10, no. 4, pp. 1514–1524, 2009.
12. K. McLean, "Evaluation of Biomaterial Performance," 2006, http://www.csiro.au/.
13. N. Ivanova, "Biomaterials: Silk as a Natural Biomaterial," 2005, http://www.sinc.sunysb.edu.edu/.
14. B. D. Ratner, A. S. Hoffmann, F. J. Schoen, and J. C. Lemons, Biomaterials Science: An Introduction to Materials in Medicine, Elsevier Academic Press, London, UK, 2nd edition, 2004.
15. P. Heffernan, "Section 3: discovery research measure (2007–2013), marine biodiscovery/biotechnology research programme," in Sea Change—A Marine Knowledge, Research and Innovation Strategy for Ireland (2007–2013), Marine Institute, Rinville Oranmore, Galway, Ireland, 2006.
16. ASMI, "Overview of biomaterials and their use in medical devices," in Handbook of Materials for Medical Devices, 2003, http://asmiinternational.org/bookstore.
17. P. Fratzl, "Biomimetic materials research: what can we really learn from nature's structural materials?" Journal of the Royal Society Interface, vol. 4, no. 15, pp. 637–642, 2007.

18. R. Schoental, "Toxicology of natural products," Food and Cosmetics Toxicology, vol. 3, no. 4, pp. 609–620, 1965.
19. G. A. Holt and A. Chandra, "Herbs in the modern healthcare environment—an overview of uses, legalities, and the role of the healthcare professional," Clinical Research and Regulatory Affairs, vol. 19, no. 1, pp. 83–107, 2002.
20. C. B. Spainhour, "Natural products," in Drug Discovery Handbook, pp. 11–72, John Wiley & Sons, Inc., New York, NY, USA, 2005.
21. B. B. Jarvis, "Evolution of metabolic pathways," in The Role of Natural Products in Evolution, J. T. Romeo, Ed., vol. 34, pp. 1–24, 1st edition, 2000.
22. http://www.thefreedictionary.com/Natural+products+chemistry.
23. http://en.wikipedia.org/wiki/Natural_product.
24. K. Nakanishi, "An historical perspective of natural products chemistry," in Comprehensive Natural Products Chemistry, vol. 8, pp. 21–48, 1999.
25. B. R. O'Keefe, "Biologically active proteins from natural product extracts," Journal of Natural Products, vol. 64, no. 10, pp. 1373–1381, 2001.
26. J. D. Connolly, "Natural products from around the world," Revista Latinoamericana de Quimica, vol. 25, no. 2, pp. 77–85, 1997.
27. K. H. Lee, "Anticancer drug design based on plant-derived natural products," Journal of Biomedical Science, vol. 6, no. 4, pp. 236–250, 1999.
28. G. W. Qin and R. S. Xu, "Recent advances on bioactive natural products from Chinese medicinal plants," Medicinal Research Reviews, vol. 18, no. 6, pp. 375–382, 1998.
29. V. J. Ram and S. Kumari, "Natural products of plant origin as anticancer agents," Drug News and Perspectives, vol. 14, no. 8, pp. 465–482, 2001.
30. C. Acharya, B. Hinz, and S. C. Kundu, "The effect of lactose-conjugated silk biomaterials on the development of fibrogenic fibroblasts," Biomaterials, vol. 29, no. 35, pp. 4665–4675, 2008.
31. C. Acharya, S. K. Ghosh, and S. C. Kundu, "Silk fibroin protein from mulberry and non-mulberry silkworms: cytotoxicity, biocompatibility and kinetics of L929 murine fibroblast

adhesion," Journal of Materials Science, vol. 19, no. 8, pp. 2827–2836, 2008.

32. N. Minoura, M. Tsukada, and M. Nagura, "Physico-chemical properties of silk fibroin membrane as a biomaterial," Biomaterials, vol. 11, no. 6, pp. 430–434, 1990. · ·

33. N. Minoura, S. I. Aiba, M. Higuchi, Y. Gotoh, M. Tsukada, and Y. Imai, "Attachment and growth of fibroblast cells on silk fibroin," Biochemical and Biophysical Research Communications, vol. 208, no. 2, pp. 511–516, 1995. · ·

34. M. Santin, A. Motta, G. Freddi, and M. Cannas, "In vitro evaluation of the inflammatory potential of the silk fibroin," Journal of Biomedical Materials Research, vol. 46, no. 3, pp. 382–389, 1999. · ·

35. L. Meinel, R. Fajardo, S. Hofmann et al., "Silk implants for the healing of critical size bone defects,"Bone, vol. 37, no. 5, pp. 688–698, 2005. · ·

36. G. M. Cragg and D. J. Newman, "Natural product drug discovery in the next millenium,"Pharmaceutical Biology, vol. 39, supplement 1, pp. 8–17, 2001.

37. A. M. Burja, B. Banaigs, E. Abou-Mansour, J. Grant Burgess, and P. C. Wright, "Marine cyanobacteria — a profile source of natural products," Tetrahedron, vol. 57, no. 46, pp. 9347–9377, 2001. · ·

38. D. J. Faulkner, "Marine natural products," Natural Product Reports, vol. 17, no. 1, pp. 7–55, 2000.

39. D. J. Faulkner, "Highlights of marine natural products chemistry (1972–1999)," Natural Product Reports, vol. 17, no. 1, pp. 1–6, 2000. · ·

40. D. J. Faulkner, "Marine natural products," Natural Product Reports, vol. 19, no. 1, pp. 1–48, 2002. · ·

41. G. M. König and A. D. Wright, "Marine natural products research: current directions and future potential," Planta Medica, vol. 62, no. 3, pp. 193–211, 1996.

42. K. Liberra and U. Lindequist, "Marine fungi—a prolific resource of biologically active natural products?" Pharmazie, vol. 50, no. 9, pp. 583–588, 1995.

43. J. K. Volkman, "Australasian research on marine natural products: chemistry, bioactivity and ecology," Marine and Freshwater Research, vol. 50, no. 8, pp. 761–779, 1999.
44. B. Haefner, "Drugs from the deep: marine natural products as drug candidates," Drug Discovery Today, vol. 8, no. 12, pp. 536–544, 2003. · ·
45. N. Fusetani, "Introduction," in Drugs from the Sea, N. Fusetani, Ed., Karger, Basel, Switzerland, 2000. ·
46. A. Kelecom, "Chemistry of marine natural products: yesterday, today and tomorrow," Anais da Academia Brasileira de Ciencias, vol. 71, no. 2, pp. 249–263, 1999.
47. L. L. Hench, "Bioceramics: from concept to clinic," Journal of the American Ceramic Society, vol. 74, no. 7, pp. 1487–1510, 1991. ·
48. L. L. Hench and S. Best, "Ceramics, glasses and glass-ceramics," in Biomaterials Science: An Introduction to Materials in Medicine, B. D. Ratner, A. S. Hoffmann, F. J. Schoen, and J. E. Lemons, Eds., Elsevier Academic Press, London, UK, 2004.
49. S. H. Gupta SH, "Mineralized tissues," Tech. Rep., Department of Biomaterials, Biomaterials, Max Planck Institute of Colloids and Interfaces, Postdam, Germany, 2005-2006.
50. P. Fratzl, "Bone material quality and Osteoporosis research," in Research in the Department of Biomaterials, Biomaterials, Max Planck Institute of Colloids and Interfaces, Biannual Report (2005-2006), P. Fratzl, Ed., Postdam, Germany, 2006.
51. P. Fratzl, The Bone Quality Book—A Guide to Factors Influencing Bone Strength, Materials Properties: Mineral Crystals, Excerota Medica, Amsterdam, The Netherlands, 2006, Edited by Dempster D., Felsenberg D. and Van Der Geest S.
52. F. C. M. Driessens, "Formation and stability of calcium phosphate in relatuion to the phase composition of the mineral in calcified tissue," in Bioceramics of Calcium Phosphate, K. DeGroot, Ed., CRC Press, Boca Raton, Fla, USA, 1983.
53. C. Balázsi, F. Wéber, Z. Kövér, E. Horváth, and C. Németh, "Preparation of calcium-phosphate bioceramics from natural resources," Journal of the European Ceramic Society, vol. 27, no. 2-3, pp. 1601–1606, 2007.

54. P. Chattopadhyay, S. Pal, A. K. Wahi, L. Singh, and A. Verma, "Synthesis of crystalline hydroxyapetite from coral (Gergonacea sp) and cytotoxicity evaluation," Trends in Biomaterials and Artificial Organs, vol. 20, no. 2, pp. 139–142, 2007.
55. G. Guillenium, J. Patat, J. Fournie, and M. Chetail, "The use of coral as a bone graft substitute,"Journal of Biomedical Materials Research, vol. 21, no. 5, pp. 557–567, 1987.
56. M. Richard, E. Aguado, G. Daculsi, and M. Cottrel, "Ultrastructural and electron diffraction of the bone-ceramic interfacial zone in coral and biphasic calcium phosphate implants." Calcified Tissue International, vol. 62, no. 5, pp. 437–442, 1998.
57. C. Demers, C. R. Hamdy, K. Corsi, F. Chellat, M. Tabrizian, and L. Yahia, "Natural coral exoskeleton as a bone graft substitute: a review," Bio-Medical Materials and Engineering, vol. 12, no. 1, pp. 15–35, 2002.
58. J. Chou, B. Ben-Nissan, A. H. Choi, R. Wuhrer, and D. Green, "Conversion of coral sand to calcium phosphate for biomedical applications," Journal of the Australian Ceramic Society, vol. 43, no. 1, pp. 44–48, 2007.
59. J. Hu, J. J. Russell, R. Vago, and B. Ben-Nissan, "Production and analysis of hydroxyapatite from Australian corals via hydrothermal process," Journal of Materials Science Letters, vol. 20, no. 1, pp. 85–87, 2001. · ·
60. R. Z. LeGeros, "Properties of osteoconductive biomaterials: calcium phosphates," Clinical Orthopaedics and Related Research, no. 395, pp. 81–98, 2002.
61. Y. Barbotteau, J. L. Irigaray, and J. F. Mathiot, "Modelling by percolation theory of the behaviour of natural coral used as bone substitute," Physics in Medicine and Biology, vol. 48, no. 21, pp. 3611–3623, 2003. · ·
62. F. Roux, D. Brasnu, B. Loty, B. Georges, and G. Guillenimin, "Madreporic coral: a new bone graft substitute for cranial surgery," Journal of Biomedical Materials Research, vol. 69, no. 4, pp. 510–513, 1988.
63. E. Arnaud, C. Morieux, M. Wybier, and M. C. De Vernejoul, "Potentiation of transforming growth factor (TGF-β1) by natural coral and fibrin in a rabbit cranioplasty model," Calcified Tissue International, vol. 54, no. 6, pp. 493–498, 1994.

64. R. A. Yukna, "Clinical evaluation of coralline calcium carbonate as a bone replacement graft material in human periodontal osseous defects," Journal of Periodontology, vol. 65, no. 2, pp. 177–185, 1994.
65. A. J. Awang-Hazmi, A. B. C. Z. Zuki, M. M. Noordim, A. Jalila, and Y. Norimah, "Mineral composition of the cockle (Anadara Granosa) shells of west coast of peninsular malaysia and it's potential as biomaterial for use in bone repair," Journal of Animal and Veterinary Advances, vol. 6, no. 5, pp. 591–594, 2007.
66. A. B. Rodríguez-Navarro, C. CabraldeMelo, N. Batista, et al., "Microstructure and crystallographic-texture of giant barnacle (Austromegabalanus psittacus) shell," Journal of Structural Biology, vol. 156, no. 2, pp. 355–362, 2006.
67. R. C. Herdman, "U.S. Congress, office of technology assessment, biopolymers: making materials nature's way," Tech. Rep. OTA-BP-E-102, US Government Printing Office, Washington, DC, USA, 1998, 1998, http://www.fas.org/ota/reports/9313.pdf.
68. D. R. Lu, C. M. Xiao, and S. J. Xu, "Starch-based completely biodegradable polymer materials," Express Polymer Letters, vol. 3, no. 6, pp. 366–375, 2009.
69. R. E. Baier, "Advanced biomaterials development from natural products," Journal of Biomaterials Applications, vol. 2, no. 4, pp. 615–626, 1988.
70. R. L. Reis, Natural-Based Polymers for Biomedical Applications, Woodhead Publishing Limited, Abinghton, UK, 2008.
71. L. G. Griffith, "Biomaterials," WTEC Panel Report on Tissue Engineering Research, WTEC, 2003.
72. T. J. Lujan, "Biomaterials science ME 491: lecture 5: natural polymers," 2010, http://www.biomechresearch.org/staff/lujan/teaching/Biomtrl_5%20Natural%20Polymers.pdf.
73. "Natural Polymers US Industry Study with Forecasts to 2010 and 2015," 2007, http://www.bharatbook.com.
74. S. L. Cooper, S. A. Visser, R. W. Hergenrother, and N. M. K. Lamba, "Polymer," in Biomaterials Science: An Introduction to Materials in Medicine, B. D. Ratner, A. S. Hoffman, F. J. Schoen, and J. E. Lemons, Eds., pp. 67–79, Elsevier, London, UK, 2nd edition, 2004.

75. I. V. Yannas, "Natural Materials," in Biomaterials Science: An Introduction to Materials in Medicine, B. D. Ratner, A. S. Hoffman, F. J. Schoen, and J. E. Lemons, Eds., Elsevier, London, UK, 2nd edition, 2004.
76. P. R. Shewry, A. S. Tatham, and A. J. Bailey, Elastomeric Proteins: Structures, Biomechanical Properties, and Biological Roles, Cambridge University Press, 2003.
77. K. Yao, J. Mao, J. Yin, et al., "Chitosan/gelatin network based biomaterials in tissue engineering," Biomedical Engineering Applications Basis and Communications, vol. 14, no. 3, pp. 115–121, 2002. ·
78. S. B. Lee, H. W. Jeon, Y. W. Lee, et al., "Artificial dermis composed of gelatin, hyaluronic acid and (1–3), (1–6), -β—Glucan," Macromolecular Research, vol. 11, no. 5, pp. 368–374, 2003.
79. J. Wood, "Tissue-like constructs made in minutes," Materials Today, vol. 8, no. 12, p. 22, 2003.
80. H. J. Jin, J. Park, R. Valluzzi, P. Cebe, and D. L. Kaplan, "Biomaterial films of bombyx mori silk fibroin with poly (ethylene oxide)," Biomacromolecules, vol. 5, no. 3, pp. 711–717, 2004. ·
81. S. Prasong, S. Yaowalak, and S. Wilaiwan, "Characteristics of silk fiber with and without sericin component: a comparison between botnbyx tnori and philosamia ricini silks," Pakistan Journal of Biological Sciences, vol. 12, no. 11, pp. 872–876, 2009. · ·
82. A. Mueller, Engineering Biomedical Materials from Corn and Wool, Alcht, Chemical Engineering Progress (CEP), 2005.
83. T. Aboushwareb, D. Eberli, C. Ward, et al., "A keratin biomaterial gel hemostat derived from human hair: evaluation in a rabbit model of lethal liver injury," Journal of Biomedical Materials Research, vol. 90, no. 1, pp. 45–54, 2009. ·
84. P. Sierpinski, J. Garrett, J. Ma, et al., "The use of keratin biomaterials derived from human hair for the promotion of rapid regeneration of peripheral nerves," Biomaterials, vol. 29, no. 1, pp. 118–128, 2008. · ·
85. W. F. P. Neves-Junior, M. Ferreira, M. C. O. Alves, et al., "Influence of fabrication process on the final properties of natural-rubber latex tubes for vascular prosthesis," Brazilian Journal of Physics, vol. 36, no. 2, pp. 586–591, 2006.

86. F. Mrue, J. Coutinho-Netto, R. Ceneviva, J. J. Lachat, J. A. Thomazini, and H. Tambelini, "Evaluation of the biocompatibility of a new biomembrane," Materials Research, vol. 7, no. 2, pp. 277–283, 2004.
87. E. S. Stevens, Green Plastics, an Introduction to the New Acience of Biodegradable Plastics, Princeton University Press, Princeton, NJ, USA, 2002.
88. A. Atala and D. J. Mooney, Synthetic Biodegradable Polymer Scaffolds, Birkhäuser, Boston, Mass, USA, 1997.
89. A. R. Talja, Preparation and characterization of potato starch films plasticized with polyols, Ph.D. thesis, University of Helsinki, Helsinki, Finland, 2007.
90. A. K. Sugih, Synthesis and properties of starch based biomaterials, Ph.D. thesis, University of Groningen, Groningen, The Netherlands, 2008.
91. C. G. Biliaderis, "Structure and phase transition of starch in food systems," Food Technology, vol. 46, no. 6, pp. 98–109, 1992.
92. J. C. Vengal and M. Srikumar, "Processing and study of novel lignin-starch and lignin-gelatin biodegradable polymeric films," Trends in Biomaterials and Artificial Organs, vol. 18, no. 2, pp. 237–241, 2005.
93. L. S. Nair and C. T. Laurencin, "Polymers as biomaterials for tissue engineering and controlled drug delivery," Advances in Biochemical Engineering/Biotechnology, vol. 102, pp. 47–90, 2006.
94. A. P. Marques, Biofunctionality and immunocompatibility of starch-based biomaterials, Ph.D. thesis, Departmento engenharia polimerios, Universidale Do Minho, Escola de Engenharia, 2004.
95. K. Sriroth, R. Onollakup, K. Piyachomkwan, and C. G. Oates, "Biodegradable plastics from cassava starch in Thailand," 1999, http://www.ciat.cgiar.org/Paginas/index.aspx.
96. E. Khor, "Chitin: a biomaterial in waiting," Current Opinion in Solid State and Materials Science., vol. 6, no. 4, pp. 313–317, 2002.
97. E. Khor, "Chitin and chitosan as biomaterials: going forward based on lessons learnt," Journal of Metals, Materials and Minerals, vol. 15, no. 1, pp. 69–72, 2005.

98. J. F. Louvier-Hernandez, R. A. Mauricio-Sanchez, G. A. Camacho-Bragado, G. Luna-Barcenas, and R. B. Gupta, "Chitin nanofibrous three-dimensional scaffold prepared by supercritical antisolvent precipitation," in Proceedings of the 8th Inter American Congress of Electron Microscopy, La Hanaba, Cuba, September 2005.
99. R. Naznin, "Extraction of chitin and chitosan from shrimp (Metapenaeus monoceros) shell by chemical method," Pakistan Journal of Biological Sciences, vol. 8, no. 7, pp. 1051–1054, 2005.
100. P. K. Dutta, J. Duta, and V. S. Tripathi, "Chitin and chitosan: chemistry, properties and applications,"Journal of Scientific and Industrial Research, vol. 63, no. 1, pp. 20–31, 2004.
101. M. M. Saad, "Chelating ability of the chitosan-glucan complex from aspergillus niger NRRL595 biomass recycling in citric acid production," Research Journal of Agric & Biological Sciences, vol. 2, no. 3, pp. 132–136, 2006.
102. S. Hirano and T. Midorikawa, "Novel method for the preparation of N-acylchitosan fiber and N-acylchitosan-cellulose fiber," Biomaterials, vol. 19, no. 1–3, pp. 293–297, 1998.
103. A. Blasinka, I. Krucinska, and M. Chrzanowski, "Dibutyrylchitin nonwoven biomaterials manufacturing using electrospinning method," Fibres and Textiles in Eastern Europe, vol. 4, no. 48, pp. 51–55, 2004.
104. R. Shelma, W. Paul, and C. P. Sharma, "Chitin nanofibre reinforced thin chitosan films for wound healing application," Trends in Biomaterials and Artificial Organs, vol. 22, no. 2, pp. 107–111, 2008.
105. M. Beran, M. Urban, L. Adamek, L. Jandusik, and J. Speracek, "Applications of mushroom chitosans in medical biomaterials," 2007, http://www.vupp.cz/.
106. S. Honary, M. Maleki, and M. Karami, "The effect of chitosan molecular weight on the properties of alginate/chitosan microparticles containing prednisolone," Tropical Journal of Pharmaceutical Research, vol. 8, no. 1, pp. 53–61, 2009.
107. "Natural polymers US industry study with forecasts to 2010 and 2015," 2007,http://www.bharatbook.com.

108. Y. M. Tsujisaka and M. Tsuhashi, "Pullulan," in Industrial Gums: Polysaccharides and Their Derivatives, R. Whistkler and J. N. BeMiller, Eds., pp. 447–460, Academic Press, San Diego, Calif, USA, 1993.

109. M. R. Rekha and C. P. Sharma, "Pullulan as a promising biomaterials for biomedical applications: a perspective," Trends in Biomaterials and Artificial Organs, vol. 20, no. 2, pp. 116–121, 2007.

110. US Congress, "Office of technology assessment, biopolymers: making materials nature's way," Tech. Rep. OTA-BP-E-102, US Government Printing Office, Washington, DC, USA, 1993.

111. U. Hasegawa, S. M. Nomura, S. C. Kaul, T. Hirano, and K. Akiyosh, "Nanogel-quantum dot hybrid nanoparticles for live cell imaging," Biochemical and Biophysical Research Communications, vol. 331, no. 4, pp. 917–921, 2005.

112. H. Hasuda, O. H. Kwon, I. K. Kang, and Y. Ito, "Synthesis of photoreactive pullulan for surface modification," Biomaterials, vol. 26, no. 15, pp. 2401–2406, 2005.

113. Y. Ito and M. Nogawa, "Preparation of a protein micro-array using a photo-reactive polymer for a cell-adhesion assay," Biomaterials, vol. 24, no. 18, pp. 3021–3026, 2003.

114. D. Hull and T. W. Clyne, An Introduction to Composite Materials, Cambridge University Press, Cambridge, Mass, USA, 2nd edition, 1981.

115. F. L. Matthews and R. D. Rawlings, Composite Materials Engineering and Science, Chapman & Hall/CRC, London, UK, 1994.

116. O. Gåserød, A. Sannes, and G. Skjåk-Bræk, "Microcapsules of alginate-chitosan. II. a study of capsule stability and permeability," Biomaterials, vol. 20, no. 8, pp. 773–783, 1999.

117. A. Sannino, S. Pappadà, M. Madaghiele, A. Maffezzoli, L. Ambrosio, and L. Nicolais, "Crosslinking of cellulose derivatives and hyaluronic acid with water-soluble carbodiimide," Polymer, vol. 46, no. 21, pp. 11206–11212, 2005.

118. T. Ito, Y. Yeo, C. B. Highley, E. Bellas, C. A. Benitez, and D. S. Kohane, "The prevention of peritoneal adhesions by in situ cross-linking hydrogels of hyaluronic acid and cellulose derivatives,"Biomaterials, vol. 28, no. 6, pp. 975–983, 2007.

119. A. Sannino, C. Demitri, and M. Madaghiele, "Biodegradable cellulose-based hydrogels: design and applications," Materials, vol. 2, no. 2, pp. 353–373, 2009.
120. Y. Tokiwa, A. Iwamoto, and M. Koyama, "Development of biodegradable plastics containing polycaprolactose and/or starch," American Chemical Society, Division of Polymeric Materials, vol. 63, pp. 742–746, 1990.
121. Y. Tokiwa, T. Ando, T. Suzuki, and T. Takeda, "Biodegradation of synthetic polymers containing ester bonds," Proceedings of the ACS Division of Polymeric Materials, vol. 62, pp. 988–992, 1990.
122. M. Zhai, L. Zhao, F. Yoshii, and T. Kume, "Study on antibacterial starch/chitosan blend film formed under the action of irradiation," Carbohydrate Polymers, vol. 57, no. 1, pp. 83–88, 2004.
123. D. R. Lu, C. M. Xiao, and S. J. Xu, "Starch-based completely biodegradable polymer materials," Express Polymer Letters, vol. 3, no. 6, pp. 366–375, 2009.
124. S. C. Mendes, R. L. Reis, Y. P. Bovell, A. M. Cunha, C. A. Van Blitterswijk, and J. D. De Bruijn, "Biocompatibility testing of novel starch-based materials with potential application in orthopaedic surgery: a preliminary study," Biomaterials, vol. 22, no. 14, pp. 2057–2064, 2001.
125. A. P. Marques, R. L. Reis, and J. A. Hunt, "The biocompatibility of novel starch-based polymers and composites: in vitro studies," Biomaterials, vol. 23, no. 6, pp. 1471–1478, 2002.
126. H. S. Azevedo, F. M. Gama, and R. L. Reis, "In vitro assessment of the enzymatic degradation of several starch based biomaterials," Biomacromolecules, vol. 4, no. 6, pp. 1703–1712, 2003.
127. L. F. Boesel, J. F. Mano, and R. L. Reis, "Optimization of the formulation and mechanical properties of starch based partially degradable bone cements," Journal of Materials Science, vol. 15, no. 1, pp. 73–83, 2004.
128. A. V. Reis, M. R. Guilherme, T. A. Moia, L. H. C. Mattoso, E. C. Muniz, and E. B. Tambourgi, "Synthesis and characterization of a starch-modified hydrogel as potential carrier for drug delivery system," Journal of Polymer Science, A, vol. 46, no. 7, pp. 2567–2574, 2008.

129. N. A. Peppas, P. Bures, W. Leobandung, and H. Ichikawa, "Hydrogels in pharmaceutical formulations," European Journal of Pharmaceutics and Biopharmaceutics, vol. 50, no. 1, pp. 27–46, 2000.
130. K. J. Zhu, L. Xiangzhou, and Y. Shilin, "Preparation, characterization and properties of polylactide (PLA)-poly(ethylene glycol) (PEG) copolymers. A potential drug carrier," Journal of Applied Polymer Science, vol. 39, no. 1, pp. 1–9, 1990.
131. C. S. Pereira, A. M. Cunha, R. L. Reis, B. Vázquez, and J. San Román, "New starch-based thermoplastic hydrogels for use as bone cements or drug-delivery carriers," Journal of Materials Science, vol. 9, no. 12, pp. 825–833, 1998.
132. C. Bastiol, "Starch-polymer composites," in Degradable Polymer-Principles and Applications, p. 112, Chapman & Hall, London, UK, 1995.
133. R. L. Reis, A. M. Cunha, and M. J. Bevis, "Using non-conventional processing routes to develop anisotropic and buiodegradable composites of starch-based thermoplastic reinforced with bone-like ceramics," Medical Plastics and Biomaterials Magazine, vol. 4, pp. 46–50, 1997.
134. W. Bonfield, M. D. Grynpas, A. E. Tully, J. Bowman, and J. Abram, "Hydroxyapatite reinforced polyethylene—a mechanically compatible implant material for bone replacement," Biomaterials, vol. 2, no. 3, pp. 185–186, 1981.
135. W. Bonfield, "Composite biomaterials, in bioceramics," in Proceedings of the 9th International Symposium on Ceramics in Medicine, T. Kukubo, T. Nakamura, and F. Miyaji, Eds., Pergamon, Otsu, Japan, 1996.
136. C. L. Jardelino, D. R. Gomes, I. I. Castro-Silva, M. H. Rocha-Leao, A. M. Rossi, and J. M. Granjeiro, "Tissue behaviour in response to alginate-hydroxyapatite-capsule containing membrane," inProceedings of the 11th International Conference on Advanced Materials (ICAM ‹09), Rio de Janeiro, Brazil, September0 2009.
137. T. W. Bauer and G. F. Muschler, "Bone graft materials: an overview of the basic science," Clinical Orthopaedics and Related Research, vol. 371, pp. 10–27, 2000.

138. W. K. Czaja, D. J. Young, M. Kawecki, and R. M. Brown, "The future prospects of microbial cellulose in biomedical applications," Biomacromolecules, vol. 8, no. 1, pp. 1–12, 2007. · ·
139. Y. Levi-Kalisman, G. Falini, L. Addadi, and S. Weiner, "Structure of the nacreous organic matrix of a bivalve mollusk shell examined in the hydrated state using Cryo-TEM," Journal of Structural Biology, vol. 135, no. 1, pp. 8–17, 2001.
140. A. G. Checa, J. H. E. Cartwright, and M. G. Willinger, "The key role of the surface membrane in why gastropod nacre grows in towers," Proceedings of the National Academy of Sciences of the United States of America, vol. 106, no. 1, pp. 38–43, 2009. · ·
141. S. Camprasse, G. Camprasse, M. Pouzol, and E. Lopez, "Artifical dental root made of natural calcium carbonate (bioracine)," Clinical Materials, vol. 5, no. 2–4, pp. 235–250, 1990. ·
142. G. Atlan, N. Balmain, S. Berland, B. Vldal, and E. Lopez, "Reconstruction of human maxillary defects with nacre powder: histological evidence for bone regeneration," Comptes Rendus de l›Academie des Sciences—Serie III, vol. 320, no. 3, pp. 253–258, 1997. ·
143. M. Joshi, S. W. Ali, and R. Purwar, "Ecofriendly antimicrobial finishing of textiles using bioactive agents based on natural products," Indian Journal of Fibre and Textile Research, vol. 34, no. 3, pp. 295–304, 2009.
144. V. R. Sinha and R. Kumria, "Polysaccharides in colon-specific drug delivery," International Journal of Pharmaceutics, vol. 224, no. 1-2, pp. 19–38, 2001.
145. M. Veerapandian and K. Yun, "The state of the art in biomaterials as nanobiopharmaceuticals," Digest Journal of Nanomaterials and Biostructures, vol. 4, no. 2, pp. 243–262, 2009.
146. F. Iemma, U. G. Spizzirri, F. Puoci, et al., "pH-Sensitive hydrogels based on bovine serum albumin for oral drug delivery," International Journal of Pharmaceutics, vol. 312, no. 1-2, pp. 151–157, 2006. ·
147. C. Y. Gan, L. H. Cheng, E. T. Phuah, P. N. Chin, A. F. M. AlKarkhi, and A. M. Easa, "Combined cross-linking treatments of bovine serum albumin gel beadlets for controlled-delivery of caffeine," Food Hydrocolloids, vol. 23, no. 5, pp. 1398–1405,

2009.

148. O. A. Diak, A. Bani-Jaber, B. Amro, D. Jones, and G. P. Andrews, "The manufacture and characterization of casein films as novel tablet coatings," Food and Bioproducts Processing C, vol. 85, no. 3, pp. 284–290, 2007.

149. K. Yamauchi, H. Hojo, Y. Yamamoto, and T. Tanabe, "Enhanced cell adhesion on RGDS-carrying keratin film," Materials Science and Engineering C, vol. 23, no. 4, pp. 467–472, 2003.

Citations

CHAPTER 1

Venkatesh Kodur, "Properties of Concrete at Elevated Temperatures," ISRN Civil Engineering, vol. 2014, Article ID 468510, 15 pages, 2014 doi:10.1155/2014/468510.

CHAPTER 2

Prakash C. Thapliyal and Kirti Singh, "Aerogels as Promising Thermal Insulating Materials: An Overview,"Journal of Materials, vol. 2014, Article ID 127049, 10 pages, 2014. doi:10.1155/2014/127049.

CHAPTER 3

C. I. Elsner, P. R. Seré, and A. R. Di Sarli, "Atmospheric Corrosion of Painted Galvanized and 55%Al-Zn Steel Sheets: Results of 12 Years of Exposure," International Journal of Corrosion, vol. 2012, Article ID 419640, 16 pages, 2012. doi:10.1155/2012/419640.

CHAPTER 4

Nurxat Nuraje, Shifath I. Khan, Heath Misak, and Ramazan Asmatulu, "The Addition of Graphene to Polymer Coatings for Improved Weathering," ISRN Polymer Science, vol. 2013, Article ID 514617, 8 pages, 2013. doi:10.1155/2013/514617.

CHAPTER 5

Y. Şahin, "Recent Progress in Processing of Tungsten Heavy Alloys," Journal of Powder Technology, vol. 2014, Article ID 764306, 22 pages, 2014, doi:10.1155/2014/764306.

CHAPTER 6

Alfonso Salinas, Maricela Lizcano, and Karen Lozano, "Synthesis of β-SiC Fine Fibers by the Forcespinning Method with Microwave Irradiation," Journal of Ceramics, vol. 2015, Article ID 217931, 5 pages, 2015. doi:10.1155/2015/217931.

CHAPTER 7

M. Antunes, V. Realinho, and J. I. Velasco, "Foaming Behaviour, Structure, and Properties of Polypropylene Nanocomposites Foams," Journal of Nanomaterials, vol. 2010, Article ID 306384, 11 pages, 2010. doi:10.1155/2010/306384.

CHAPTER 8

Oladeji O. Ige, Lasisi E. Umoru, and Sunday Aribo, "Natural Products: A Minefield of Biomaterials," ISRN Materials Science, vol. 2012, Article ID 983062, 20 pages, 2012. doi:10.5402/2012/983062.

Index

A

Acoustically induced optical Kerr effects (AIOKE) 60
Adiabatic shear banding (ASB) 159
Atomic force microscopy (AFM) 110, 113
Attenuated total reflectance (ATR) 115
Azodicarbonamide (ADC) 205

B

Body-centered cubic (BCC) 131

C

Cellulose nanofibril (CNF) 60
Chiral mesoporous SiO2 (CMS) 60
Colloidal suspension 51
Conventional furnace (CF) 142

D

Deoxyribonucleic acid (DNA) 252, 256
Depleted uranium (DU) 130, 157
Differential scanning calorimetry (DSC) 5, 210
Differential thermal analyzer (DTA) 5
Dilatometric curve 7
Dynamic mechanical analysis (DMA) 211

E

EDS (energy dispersive spectroscopy) 191
Electrical conductivity 204, 206, 207, 211, 212, 226, 227, 228, 229, 230, 233

Extracellular matrix (ECM) 250, 267

F

FESEM (field emission scanning electron microscope) 191
Fiber-reinforced concrete (FRC) 4
Forcespinning (FS) 191
Foreign material 244, 250
Fourier transform infrared (FTIR) 110, 113
Fourier transform infrared spectroscopy (FTIR) 190
FTIR (Fourier transform infrared spectroscopy) 191

G

Glass fiber-reinforced plastic (GFRP) 113

H

High melt strength polypropylenes (HMS-PP) 205
High strength concrete (HSC) 2
High-strength concrete (HSC) 4

K

kinetic energy (KE) 132, 170

L

Liquid-phase sintering (LPS) 131

M

Mechanical alloying (MA) 136
Mechanically alloyed (MAed) 136, 139
Melt flow index (MFI) 207
Microwave sintering (MW) 130
Multilayered insulation (MLI) 57

N

Normal strength concrete (NSC) 2

Normal-strength concrete (NSC) 4

O

Oxide-dispersion strengthened (ODS) 136

P

Polycarbomethylsilane (PCmS) 190, 192
Polystyrene (PS) 190, 192
Polytetrafluoroethylene (PTFE) 192
Polyurethane (PU) 110
Powder injection molding (PIM) 129, 135
Powder metallurgy (PM) 129, 171
Pulse plasma sintering (PPS) 145

R

Reinforced concrete (RC) 3

S

Saturated Calomel Electrode (SCE) 79
Scanning electron microscopy (SEM) 190
Slower strength degradation 2
Smooth muscle cells (SMCs) 267
Spark plasma sintering (SPS) 144
Spark-plasma sintering (SPS) 130, 171

T

Thermal expansion 7, 21, 22, 23, 24
Time of wetness (TOW) 80
Translucent nature 45

U

Ultimate tensile strength (UTS) 150
Ultrahigh performance concrete (UHP) 4
Ultraviolet (UV) 110

W

Water sorption 265
Wide angle X-ray scattering (WAXS) 211
W-Ni-Fe-Co (WNFC) 146
W-Ni-Fe (WNF) 146

X

X-ray diffraction (XRD) 190
XRD (X-ray diffraction) 191